Digital Image Watermarking

Theoretical and Computational Advances

Intelligent Signal Processing and Data Analysis

Series Editor
Dr. Nilanjan Dey
Department of Information Technology,
Techno India College of Technology, West Bengal, India

A Beginner's Guide to Image Pre-processing Techniques
Jyotismita Chaki, Nilanjan Dey

Bio-Inspired Algorithms in PID Controller Optimization
Jagatheesan Kallannan, Anand Baskaran, Nilanjan Dey, Amira S. Ashour

For more information about this series, please visit: https://www.crcpress.com/Intelligent-Signal-Processing-and-Data-Analysis/book-series/INSPDA

Digital Image Watermarking

Theoretical and Computational Advances

By
Surekha Borra, Rohit Thanki, and
Nilanjan Dey

CRC Press is an imprint of the
Taylor & Francis Group, an **informa** business

CRC Press
Taylor & Francis Group
6000 Broken Sound Parkway NW, Suite 300
Boca Raton, FL 33487-2742

First issued in paperback 2020

ISBN 13: 978-0-367-67035-1 (pbk)
ISBN 13: 978-1-1383-9063-8 (hbk)

Visit the Taylor & Francis Web site at
http://www.taylorandfrancis.com

and the CRC Press Web site at
http://www.crcpress.com

Contents

Preface ix

Authors xi

Abbreviations xiii

Chapter 1 ▪ Introduction 1

1.1 DIGITAL RIGHTS MANAGEMENT 2

1.2 DIGITAL IMAGE WATERMARKING 4

1.3 CLASSIFICATION OF COPYRIGHT MARKING 5

1.4 GENERAL FRAMEWORK OF DIGITAL WATERMARKING 11

1.5 PERFORMANCE CRITERIA 12

1.6 PERFORMANCE METRICS 15

1.7 ATTACKS ON WATERMARKS 16

1.8 DIGITAL IMAGE WATERMARKING TOOLS 19

REFERENCES 22

Chapter 2 ▪ Advanced Watermarking Techniques 25

2.1 INTRODUCTION 26

2.2 WATERMARKING IN THE SPATIAL DOMAIN 28

2.2.1 Least-Significant-Bit Substitution Technique 29

2.2.2 Patchwork Technique 29

2.2.3 Texture Mapping Coding Technique 30
2.2.4 Predictive Coding Technique 30
2.2.5 Additive Watermarking Technique 31
2.2.6 Other Spatial Domain Watermarking Techniques 32
2.3 WATERMARKING IN THE TRANSFORM DOMAIN 33
2.4 WATERMARKING IN THE DISCRETE COSINE TRANSFORM 34
2.5 WATERMARKING IN THE DISCRETE WAVELET TRANSFORM 37
2.6 WATERMARKING USING SINGULAR VALUE DECOMPOSITION 38
2.7 COMPRESSIVE SENSING AND QR DECOMPOSITION METHODS 40
2.8 SCHUR DECOMPOSITION–BASED WATERMARKING 41
2.9 HESSENBERG MATRIX FACTORIZATION IN WATERMARKING 41
2.10 VISIBLE AND REVERSIBLE WATERMARKING 42
2.11 MACHINE LEARNING–BASED IMAGE WATERMARKING 44
2.12 CHALLENGES 57
REFERENCES 57

CHAPTER 3 ▪ Watermarking Using Bio-Inspired Algorithms 71

3.1 OPTIMIZATION AND ITS APPLICATION TO DIGITAL IMAGE WATERMARKING 72
3.2 IMAGE WATERMARKING USING GENETIC ALGORITHM (GA) AND GENETIC PROGRAMMING 74

3.3 IMAGE WATERMARKING USING DIFFERENTIAL EVOLUTION (DE) 80

3.4 IMAGE WATERMARKING USING SWARM ALGORITHMS 81

3.4.1 Image Watermarking Using Ant Colony and Bee Colony 84

3.4.2 Image Watermarking Using Cuckoo Search Algorithm 85

3.4.3 Image Watermarking Using Particle Swarm Optimization 88

3.4.4 Image Watermarking Using Firefly Algorithm 90

3.5 IMAGE WATERMARKING USING SIMULATED ANNEALING (SA) 97

3.6 IMAGE WATERMARKING USING TABU SEARCH 98

REFERENCES 99

CHAPTER 4 ■ Hardware-Based Implementation of Watermarking 107

4.1 INTRODUCTION 108

4.2 HARDWARE-BASED IMPLEMENTATION OF DIGITAL IMAGE WATERMARKING 109

4.2.1 Hardware-Based Implementation of Watermarking Using DSP Boards 111

4.2.2 Hardware-Based Implementation of Watermarking Using FPGA/ASIC Chip 112

4.2.3 Hardware–Software Co-Simulation 113

4.3 PERFORMANCE OF HARDWARE-BASED IMPLEMENTATION 116

4.4 CHALLENGES AND FUTURE DIRECTIONS 126

REFERENCES 127

Chapter 5 ■ Applied Examples and Future Prospectives 133

5.1 APPLICATIONS OF WATERMARKING 134
5.2 WATERMARKING IN TELEMEDICINE 136
5.3 ROLE OF WATERMARKING IN REMOTE-SENSING MILITARY 137
5.4 INDUSTRIAL AND MISCELLANEOUS APPLICATIONS 139
5.5 FUTURE PROSPECTIVES 139
REFERENCES 141

Chapter 6 ■ Case Study 143

6.1 EMBEDDING ALGORITHM 144
6.2 EXTRACTION ALGORITHM 146
6.3 SIMULATION RESULTS 147
6.4 MAIN FEATURES OF PROPOSED SCHEME 149
REFERENCES 154

INDEX 157

Preface

In the era of the internet, copyright protection of digital images plays an important role in web publishing, videos, online advertising, online repositories, libraries, and so forth. Digital watermarking techniques have proved to be an effective way to resolve rightful ownership by embedding a watermark visibly or invisibly in the image, in such a way that the owner is able to detect and extract it using a secret key. While there are various spatial and frequency domain watermarking techniques that have been developed in the past three decades, optimization is a commonly encountered mathematical problem in all engineering disciplines, including data security. The research on new embedding domains, hardware implementations, machine learning, and bio-inspired algorithms for image watermarking is on the rise. This book presents advanced designs and developments in image watermarking algorithms and hardware implementations with a special focus on optimizing methods.

This book introduces state-of-the-art watermarking techniques that have been developed in various domains, along with their optimization techniques and hardware implementations. The book also presents comparative analysis of more than a hundred watermarking techniques. Further, it covers the applications, difficulties, and challenges faced by such algorithms, as well as future directions for research.

The book is composed of six chapters, which accomplish the following:

- Provide a broad background of image watermarking
- Provide an overview of newly developed machine learning–based watermarking techniques in various independent and hybrid domains
- Provide an overview of optimization problems and solutions in watermarking with a special focus on bio-inspired algorithms
- Cover the hardware implementation of watermarking
- Highlight recent innovations, designs, developments, and topics of interest in existing image watermarking techniques for intellectual property (IP) protection
- Outline different applications of digital image watermarking

Surekha Borra
Rohit Thanki
Nilanjan Dey

Authors

Surekha Borra is currently a professor in the Department of Electronics and Communication Engineering and chief research coordinator of K. S. Institute of Technology, Bangalore, India. She earned her doctorate in the copyright protection of images from Jawaharlal Nehru Technological University, Hyderabad, India. Her current research interests are image and video analytics, machine learning, biometrics, biomedical signal, and remote sensing. She has filed 1 Indian patent; published 6 books, 12 book chapters, and several research papers in refereed and indexed journals; and has participated in conferences at the international level. She has received several research grants and awards from professional bodies and the Karnataka state government of India. She has received the Young Woman Achiever Award for her contribution to the copyright protection of images, the Distinguished Educator & Scholar Award for her contributions to teaching and scholarly activities, and the Woman Achiever's Award from the Institution of Engineers (India) for her prominent research and innovative contributions.

Rohit Thanki earned his PhD in multibiometric system security using the compressive sensing theory and watermarking from C. U. Shah University, Wadhwan City, Gujarat, India, in 2017. His areas of research are digital watermarking, the biometrics system, security, compressive sensing, pattern recognition, and image processing. He has published 5 books, 7 book chapters, and more than 25 research papers in refereed and indexed journals, and has participated in conferences at the international and national level. His international recognition includes professional memberships and services in refereed organizations and program committees, and being a reviewer for journals published by the Institute of Electrical and Electronics Engineers (IEEE), Elsevier, Taylor & Francis, Springer, and IGI Global.

Nilanjan Dey is an assistant professor in the Department of Information Technology at Techno India College of Technology, Kolkata, India. He was an honorary visiting scientist at Global Biomedical Technologies Inc., California, and an associated member of University of Reading, London, United Kingdom.

Dr. Dey has authored or edited more than 40 books with Elsevier, Wiley, CRC Press, Springer, and others, and has published more than 300 research articles. He is the editor-in-chief of the *International Journal of Ambient Computing and Intelligence* (IGI Global). Dr. Dey is the series co-editor of *Springer Tracts in Nature-Inspired Computing* (Springer Nature), *Advances in Ubiquitous Sensing Applications for Healthcare* (Elsevier), and *Intelligent Signal Processing and Data Analysis* (CRC Press). He is an associate editor of *IEEE Access*.

Dr. Dey's main research interests include medical imaging, machine learning, data mining, etc. He was recently awarded as one of India's top 10 most published and cited academics in the field of computer science for the period 2015–2017.

Abbreviations

ACO	ant colony optimization
ASIC	application-specific integrated circuit
BA	bee algorithm
BCR	bit correction rate
BER	bit error rate
BPNN	back-propagation neural network
C	host data
CNN	convolution neural network
CS	compressive sensing
CSA	cuckoo search algorithm
D	extraction algorithm
DCT	discrete cosine transform
DE	differential evolution
DFT	discrete Fourier transform
DL	deep learning
DSP	digital signal processor
DWT	discrete wavelet transform
E	embedding algorithm
f	fitness function
FA	firefly algorithm
FNN	feedforward neural network
FPGA	field-programmable group array
GA	genetic algorithm
GP	genetic programming
HDL	hardware description language

HNN	Hopfield neural network
HPI	host-port interface
ISA	industry standard architecture
K	secret key
LSB	least significant bit
ML	machine learning
NC	normalized correlation
NVF	noise visibility function
PN	pseudorandom noise
PSNR	peak signal-to-noise ratio
PSO	particle swarm optimization
RDWT	redundant discrete wavelet transform
ReLU	ratified linear unit
SA	simulated annealing
SS	spread spectrum
SSIM	structural similarity index measure
SVD	singular value decomposition
SVM	support vector machine
SVR	support vector regression
TS	tabu search
VLSI	very large-scale integrated
W	watermark
WPSNR	weighted PSNR

CHAPTER 1

Introduction

A VARIETY OF DIGITAL INFORMATION, for example, pictures, recordings, melodies, and essential archives, is being published or exchanged between people, organizations, and associations every second. The digital content and online transmission of data are fast, less expensive, and easy to store and process, and result in high-quality transmission and distribution. On the flip side, new security-related problems have arisen, such as to how to trust, identify, or authenticate the right owner/creator/correspondent, and how to confidentially and reliably protect the multimedia information/intellectual property (IP). With the illegal downloads, distributions, copying, and use of a variety of data, such as multimedia, web-published data, broadcast information, IP, and commercial designs, the creators/producers/authors/editors/distributors are experiencing great losses, and hence digital rights has become the need of the hour.

This chapter discusses the importance of digital rights management (DRM) and copyrighting images, and reviews the techniques defined in DRM for securing the image data and corresponding owners. Also explained are the differences in the concepts of encryption, steganography, and digital image watermarking. The broad classification of copyright marking methods, the generalized digital image watermarking framework, its performance criteria and metrics, and image watermarking tools are presented.

1.1 DIGITAL RIGHTS MANAGEMENT

To reduce the losses caused by piracy, and to limit, prevent, identify, manage, use, manipulate, distribute, deliver, and measure the illegal actions as well as technological solutions, at every stage of online communication, from data generation to consumption, a standardized set of rules, methods, and techniques are defined by the DRM systems globally. The DRM mainly deals with licensing agreements, data viewing, data access, copy protection, copy prevention, copy control, and technical protection measures when multimedia is stored or transferred across a variety of devices and networks. Many copyright laws and acts related to DRM, such as the Digital Millennium Copyright Act (DMCA), World Intellectual Property Organization Copyright Treaty (WCT), and European directive on copyright, are defined for access control of copyrighted works, though they are not globally accepted (Vellasques et al., 2010; Surekha and Swamy, 2014). The DRM defines an open standard that discusses the issues and requirements related to universal multimedia access (UMA), such as access control, which include user identification, level of access, copy control, the creation of unauthorized copies, the detection of illegal distribution and tracking, the prevention of users from modifying the content, secure storage and transmission using algorithms and protocols, renewability, and interoperability.

The DRM aims to prevent unauthorized access, copying, and redistribution of digital media by encrypting the data and not making it directly accessible (Mohanty et al., 2017). The DRM can be referred to as the extension to digital media copyrighting, though both have different objectives and regulations. The DRM recommends key-based encryption and watermarking algorithms, along with some security protocols for ensuring confidentiality, authentication, copy control, and data integrity. The key management plays a crucial role and has a challenging task in providing security as the working of algorithms is made open (Borda, 2005). Key generation, verification, secure storage and transfer, revocation, and key escrow are all part of key management. The usage of biometrics as secret keys helps in overcoming key exchange problems.

Cryptographic techniques are used to make the content unintelligible using encryption keys and algorithms, before its storage/publication/distribution (Thanki and Kothari, 2017). The person/device having knowledge of the decryption key is allowed to access and decrypt completely or partially the protected data upon checking his or her authorization and user rights. Encryption, which comes first in DRM, ensures access control and authentication of users or content. A limitation of encryption is the fragility of ciphertext, implying that it is impossible to decrypt the content if the ciphertext is modified even partially. Random access of scalable/multiresolution data is also not possible when data is encrypted. Further, once the data is converted back to its original form by authorized users for use, the cryptographic techniques fail to protect the ownership claims, creating an analog hole. In addition, high computational costs for encryption and decryption majorly limit its application in real time.

A digital signature, on the other hand, is a message-dependent data string that is appended to the original message using encryption techniques. The objective is to guarantee the data integrity and overcome attacks related to nonrepudiation. A digital signature can easily be removed and can be made invalid by changing the file content (Petrovic et al., 2006). Steganography and watermarking techniques address these problems (Langelaar et al., 2000; Thanki and Kothari, 2017; Thanki et al., 2017, 2018). Steganography is a kind of one-to-one secret communication technique that modifies a multimedia file to hide and detect the secret message, by authorized personnel only (Hartung and Kutter, 1999; Langelaar et al., 2000). While both steganography and encryption ensure confidentiality, steganography ensures that nobody has knowledge that entities are communicating in secret, and hence it is suitable for copyright marking. In contrast, watermarking has been part of one-to-many communications and is used to verify the owner of a multimedia file (Vellasques et al., 2010).

While the objective of steganography is to protect the hidden message, the objective of watermarking is to protect the host file from ownership/copy conflicts. The watermark represents the author/owner/buyer of the file (Dey and Santhi, 2017; Dey et al., 2017; Borra et al., 2017, 2018). In contrast to appending the signature at the end of the file, as in the case of a digital signature to ensure authorship, watermarking embeds watermarks in the file itself, guaranteeing authorship and data integrity (Sherekar et al., 1999). While the cryptography objective is to secure the file (confidentiality) being stored/transferred, watermarking tries to secure ownership (authenticity) of the digital file. While encryption cannot copy control once the data is decrypted, watermarking can protect and copy control data even after decryption (Borda, 2005; Sherekar et al., 1999).

1.2 DIGITAL IMAGE WATERMARKING

Images are often susceptible to theft and copyright infringement. There are many occasions where images have been stolen from websites for usage/fame/financial gain and justice did not prevail, leading to losses for the owners/inventors. Mechanisms for identifying images and protecting their owners from adversaries (Vellasques et al., 2010) are thus needed. Before the digital era, painters/photographers/organizations usually signed their art/photographs/designs with their signature, initials, or pseudonyms/print stamps/embossing seals to help identify themselves as the owners of the images, especially if the IP were to be shared. Watermarking is another way for an owner to sign his or her image and ruin the efforts of attackers. A watermark is a superimposed image/logo/text placed over an image with the intention of identifying the owner of the image. Analog watermarks that are visible and relatively easy to replicate have existed for centuries. These watermarks were first used in paper mills and later seen on currency notes and postal stamps.

Digital watermarking as part of the DRM system provides mechanisms to represent/record/hide the copyright owner/distributor/distribution chain/purchaser in the images for the purpose of providing evidence in cases of copyright protection and copyright law enforcement (Mohanty et al., 2017). Some operations involved in storing/editing/transmitting may distort, delete, or otherwise interfere with watermarks. Universal acceptance and the deployment of watermarking technology will only be possible after it reaches a satisfactory degree of maturity and after its standardization, guaranteeing a minimum level of quality with provable robustness and security levels. Considering this, it is evident that we are in the middle stage of the process of adopting watermark technology, working to construct effective watermarking systems.

1.3 CLASSIFICATION OF COPYRIGHT MARKING

Digital image copyright marking techniques can be classified into many categories, as shown in Figure 1.1. The classification is made considering a variety of parameters, such as the type of image to be protected, the type of copyright mark, the embedding domain, the perceptibility of the mark, the reversibility of the original document, the purpose/application of the copyright marking, its use, and the type of keys and data required at the time of copyright verification. The host image to be protected can be a binary/halftone image, gray/color image, medical image, or hyperspectral image based on the application. The watermark can be a random binary pattern, credit card number, picture, signature, logo of the owner, host image source, or host content-related data, all of which can identify the copyright information of the owner/company. Logo-indicating watermarks are usually binary with smooth/sharp details. They may also include some text information. Statistical watermarks are generated using different noises, such as pseudorandom noise (PN) and white Gaussian noise (WGN). The binary random sequences generated from these noises are often multiplied with watermarks to generate statistical watermarks. An alternative way to generate statistical watermark patterns is to use the patchwork method, in which the differences between pixels in the host image are compared with the reference value using a hypothesis test.

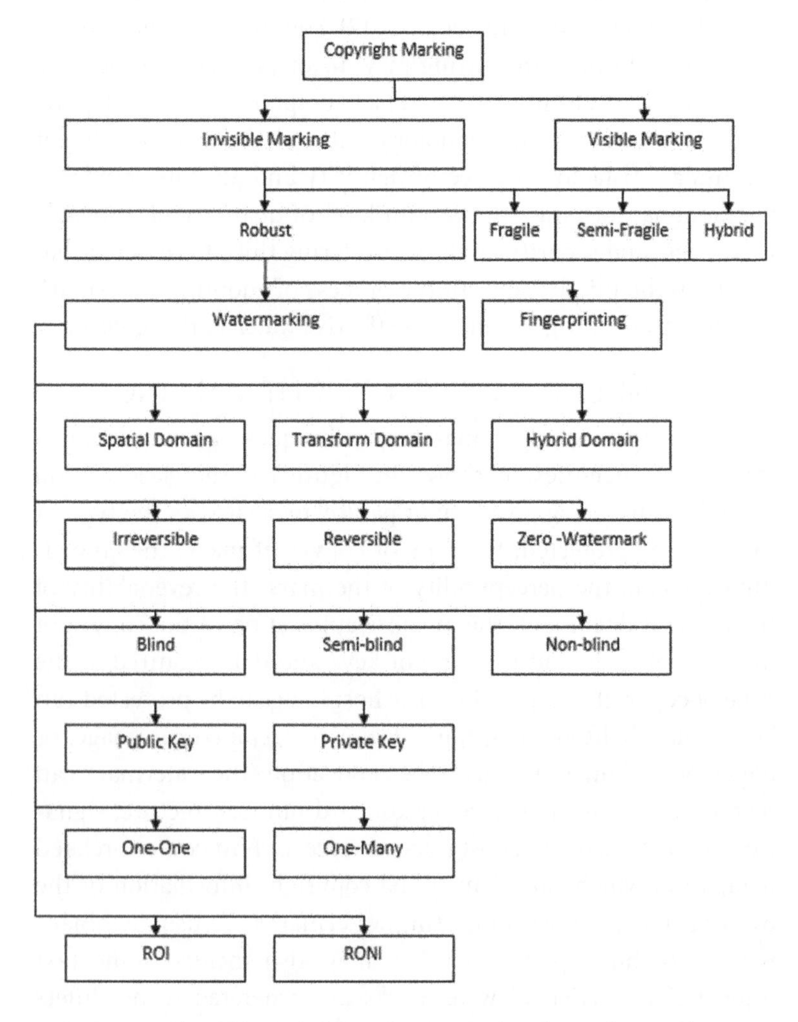

FIGURE 1.1 Classification of digital image copyright marking techniques.

Based on human perceptibility, copyright marks can be invisible/visible/dual. Visible watermarks, such as logos, are embedded transparently at image corners, warning of copyrights and ownership for content protection (Qasim et al., 2018). Placing visible watermarks at the center of an image protects against cropping but affects the quality and will not be pleasing to the viewer. Today, visible watermarks can also be placed intelligibly on the host image without affecting the quality using machine learning techniques. Invisible marks are mainly designed for forensic/investigation purposes as they are placed invisibly and randomly at any pixel positions or frequency coefficients of the image based on a secret key. They can be recovered accordingly when required for authentication, ownership verification, or data integrity validation purposes. Some applications embed both invisible and visible watermarks for improved security. There are four types of invisible marks: robust, semifragile, fragile, and hybrid techniques (Qasim et al., 2018).

Fragile marking embeds marks in imperceptible portions of an image such that the marks are destroyed in case of modification of the image. A simple way to implement fragile marking is to hide marks in the least significant bits of the image. The objective of fragile marking is to ensure that the host image is not tampered and that data integrity and authentication are maintained. Fragile marks can be easily implemented and removed, and are not suitable for proving ownership, but they are useful as evidence that an image was modified. The semifragile methods are limited in robustness and are designed to survive authorized/specific levels of image processing operations, such as compression. As these methods can differentiate some sets of operations/attacks, they are also used for checking integrity and authentication. The hybrid approaches, on the other hand, are a combination of robust and fragile methods to achieve a greater level of security, as they achieve ownership protection, data integrity, and authentication in parallel (Qasim et al., 2018).

Robust marking is preferred in cases of security applications, such as copy control, broadcast monitoring, and copyright protection as these marks survive a wide range of attacks (Qasim et al., 2018). Robust marking embeds marks in significant portions of an image such that the marks are not destroyed in cases of modifications/distortions to the image, such as compression, rescaling, resampling, analog-to-digital (A/D) and digital-to-analog (D/A) conversion, additive noise, linear and nonlinear filtering, shifting, cropping, and rotations, to name a few. The design of robust marking techniques is complex and challenging due to the trade-off among the conflicting requirements, such as hiding capacity, transparency, and robustness. Though the degree of survival of attacks depends on the application, the ideal objective is to design the marking system such that the mark can be removed only by destroying the host image, which is very useful in protecting sensitive images related to the medical field and defense (Dey et al., 2017; Biswas et al., 2013). There are two main types of robust marking based on the entity to be protected. While watermarking identifies the copyright owner of the file, fingerprinting identifies the authorized customer who allows/distributes/makes illegal copies of the host image, and/or violates the license agreements. Robust watermarking, unless specified, is irreversible. A special category called reversible watermarking is designed to be lossless as these marks are invertible and the original watermark can be restored and/or recovered. Zero watermarking, on the other hand, does not affect the quality of the host image in the process of watermark hiding.

The watermark detection itself is a function of many inputs, such as the test image, private or public key, original watermark, and original host image. Asymmetric/public key watermarking allows extraction of the watermark by anyone apart from the owner by limiting access to remove it. On the other hand, to prove ownership of images, private watermarks are used as they can be detected only by a secret key (Mohanty et al., 2017).

Blind watermarking techniques do not require either the original host or the original watermark for watermark extraction, and are flexible for use in many applications, including image authentication, copyright protection, covert communication, and electronic voting systems. Semiblind techniques do not require the original host but do require the original watermark or some additional information for watermark extraction. Content privacy, image authentication, and copyright protection are a few applications of semiblind techniques. Non-blind techniques require the original host image for watermark extraction, which may be difficult to produce, and so are limited in application.

Watermarking systems are classified into spatial and transform categories (Qasim et al., 2018) depending on the watermark embedding domain. These techniques are designed to insert the watermark directly into significant portions of the pixels/transform coefficients of the original host image if it is gray, or into the luminance or respective color components. While some watermarking techniques follow one-to-one mapping, many existing techniques are block processed to insert one bit of watermark over many pixels or transform coefficients randomly. In contrast to substitution techniques, some watermarking techniques add scaled watermark information to the host image pixels/transformed coefficients while minimizing noticeable distortions. Machine learning or optimization techniques are now in use to determine the optimum and adaptable scale factors appropriate for the host image.

The transform domain schemes are computationally complex but robust for signal processing attacks and are the right choice for resolving ownership issues. Transforms such as discrete cosine transform (DCT), discrete wavelet transform (DWT), and singular value decomposition (SVD) are widely employed due to their adaption in framing compression standards and their relative perceptual properties of frequency bands (Thanki et al., 2017). In hybrid techniques, watermark bits are inserted into hybrid coefficients of the host image, obtained by combining two or more transforms. The sparse domain techniques, on the other hand, insert a watermark into sparse measurements of the host image obtained from the compressive sensing (CS) theory.

In traditional watermarking schemes, the watermark is inserted into specified pixels or frequency coefficients of the host image using simple logical or mathematical equations that involve addition and multiplication. In additive watermarking, the watermark is inserted into the host image with the help of a gain factor, as in Equation 1.1.

$$\mathrm{WC} = C + k \times w \tag{1.1}$$

In multiplicative watermarking, the watermark is inserted using a constant and weighted factor, which in turn is multiplied by the pixel information or frequency coefficients of the host image to get the watermarked image, as in Equation 1.2.

$$\mathrm{WC} = C * (1 + k \times w) \tag{1.2}$$

Intelligent watermarking schemes, on the other hand, use various intelligent algorithms, such as machine learning, deep learning, optimization techniques, and bio-inspired algorithms, to improve the results of traditional watermarking (Dey et al., 2014). The machine learning and optimization algorithms are widely used in watermark embedding and/or the extraction process. The bio-inspired and optimization techniques, on the other hand, are used for the optimum selection of pixels or frequency coefficients of the host image for watermark embedding. These algorithms are also used for automatically finding the optimized scaling factors during watermark embedding. This is in contrast to traditional watermarking, where watermark bits are inserted into the host image using user-defined locations/scaling factors.

1.4 GENERAL FRAMEWORK OF DIGITAL WATERMARKING

Digital copyright watermarking enables us to bring copyright violators to court, as the embedded copyright mark in any legally published/sold image is retained and can be extracted in any copies made. Digital image watermarking is a process of embedding a watermark into the image so as to extract it at a later stage to detect ownership identity. Figure 1.2 shows a generalized block diagram of robust invisible image watermarking. The host is the raw digital image that has to be protected by inserting the watermark. The watermark can be a message/logo/statistical pattern inserted into the host that has some relevance to the host. The general framework of digital image watermarking and the copyright authentication process is composed of three major components: (1) the embedder,

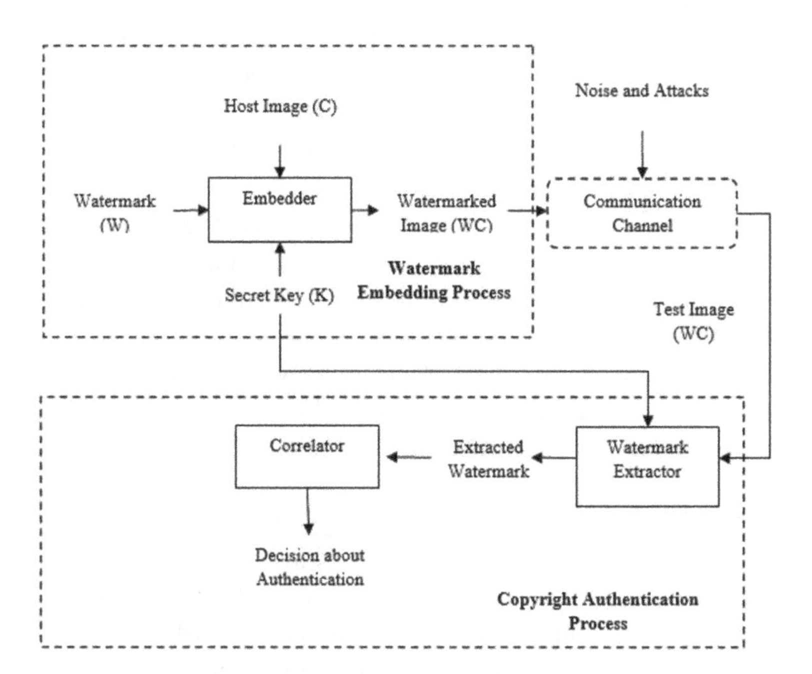

FIGURE 1.2 Generalized block diagram of digital image watermarking.

(2) the extractor, and (3) the correlator. The watermark may be encrypted before its insertion into the host image. In such cases, the extracted watermark has to be decrypted before it is compared to the original to make an assertion about ownership. All the components may be hardware units or software programs.

The embedder (E) is a function of the host image (C), secret key (K), and one or more watermarks (W). The watermark W may be a random sequence or a meaningful message or image, such as logo or copyright information. The embedder outputs a watermarked image (WC) such that $\text{WC} = E\ (C, K, W)$. The produced watermarked image (WC) can be stored, transmitted, or published. The owner must extract the watermark in order to prove ownership. The watermark extraction process may or may not be the inverse of embedding. The extractor function (D) accepts a secret key (K), test image (T), and/or the original nonwatermarked image (C) and original watermark (W) to detect the watermark (W') such that $W' = D\ (K, T, C, W)$. The test image can be the watermarked image (WC), an attacked image (AC), a nonwatermarked original image (C), or some other unauthorized image. The detected watermark (W') is correlated with the original to obtain a similarity score (Abdelhakim et al., 2018). The correlator outputs 1 if the similarity score exceeds a predefined threshold indicating that the watermark is verified and the image is authenticated. A public/private key is used in the embedding and extraction process to achieve confidentiality. The process involved in embedder/extractor and the inputs/outputs to these components vary depending on the type of watermarking and application.

1.5 PERFORMANCE CRITERIA

The performance of image watermarking and its evaluation depend on several factors, such as the type of host image and application. The subjective and objective analysis helps in identifying how well the watermark is hidden in the host image without being perceived by the human eye. The general performance criteria and essential requirements of any invisible robust image watermarking technique are given below (Mohanty et al., 2017; Qasim et al., 2018):

- Imperceptibility/fidelity: The watermark, once inserted into the host image, must be perceptually indiscernible. Imperceptibility is a measure of the perceptual transparency of a watermark and is an important requirement of invisible watermarking. It is required that the watermarked image statistically similar to its original. Imperceptibility often conflicts with watermark size and robustness. The similarity of the watermarked and original/attacked images is usually calculated by metrics such as the structural similarity index measure (SSIM) and peak signal-to-noise ratio (PSNR) (Kutter and Petitcolas, 1999; Wang and Bovik 2002). In some specific applications, it is required that there not be any degradation of quality of the host image, which makes the design of a digital watermark extremely difficult.
- Security: This refers to the detectability and key restrictions. The watermarking scheme must be secure against the unauthorized detection and modification of the embedded watermark by attackers or imposters who have knowledge of embedding and extracting algorithms. A digital watermark must be secure enough. It should be difficult for the adversary to remove even partial information of watermark without destroying the cover image.
- Robustness: A watermarked image undergoes unintentional transformations, such as compression during storing and transmission, and intentional attacks, such as cryptographic, removal, resampling, cropping, geometric, and scaling attacks. The resistance of a digital watermark to unforeseen and designated attacks is crucial in watermarking and is mainly dependent on the embedding domain and selection of pixels/coefficients. In practice, watermarking algorithms

cannot survive all possible attacks. Note that the level of robustness required depends on the type of application.

- Capacity/data payload: This is the number of bits a watermarking scheme can insert into a host image without affecting its quality and robustness. The capacity requirements are application oriented. Large payloads reduce the probability of coincidence but allow easy tampering and have a high impact on imperceptibility and robustness.
- Computational complexity: This is the time taken to embed/extract the watermark. While high-security applications demand high computational complexity, real-time applications need faster algorithms. In practice, there should be seamless overhead during watermark implementation, extraction, and verification in terms of cost and time.
- Reliability: This is achieved with authentication and data integrity. The ability to identify host image origin/owner is referred to as authentication. Data integrity, on the other hand, ensures that the watermark is not modified by unauthorized entities.
- False-positive rate/probability of coincidence: A false-positive arises if a watermark is detected from nonwatermarked images. Large watermarks result in a smaller false-positive rate.
- Cost: The design cost with respect to area, power, and resources must be minimal when watermarking is implemented in hardware.
- Reversibility: In specific applications such as the medical field, slight modifications to images could lead to disaster, which may include legal implications. In such cases, it is necessary to strictly retrieve the original host image when required. These applications demand the development of lossless or reversible watermarking techniques, where recovery of the original image is possible after proving ownership.

1.6 PERFORMANCE METRICS

The performance of image watermarking schemes in terms of imperceptibility and robustness can be measured by various quality measures. To measure the similarity between the original and watermarked image for perceptual distortions, imperceptibility metrics such as PSNR and weighted PSNR (WPSNR) (Kutter and Petitcolas, 1999; Thanki et al., 2018) are used. The normalized correlation (NC) and SSIM are commonly used metrics for the measurement of similarity between the original and extracted watermarks at the correlator. These measures indicate the quality of the extracted watermark and also help analyze the robustness of the watermarking technique in the presence of attacks (Thanki et al., 2017).

The PSNR given in Equation 1.3 is measured in decibels and depends on the mean square error (MSE), which is an error between the original and watermarked image. The MSE is calculated using Equation 1.4:

$$\mathrm{PSNR} = 10 \times \log_{10}\left(\frac{255^2}{\mathrm{MSE}}\right) \tag{1.3}$$

$$\mathrm{MSE} = \frac{1}{M \times N}\sum_{x=0}^{M-1}\sum_{y=0}^{N-1}\left(C(x,y) - \mathrm{WC}(x,y)\right)^2 \tag{1.4}$$

where C is the original host image and WC is the watermarked image, respectively.

The WPSNR is a new approach for calculating the imperceptibility of watermarked images and is shown in Equation 1.5:

$$\mathrm{WPSNR} = 10 \times \log_{10}\left(\frac{255^2}{\mathrm{NVF} \times \mathrm{MSE}}\right) \tag{1.5}$$

where NVF is the noise visibility function, which represents the texture information of an image based on the Gaussian model. The value of NVF lies in the interval [0, 1]. The NVF value is 0 for the texture region and 1 for the flat region. The NVF is calculated using Equation 1.6:

$$\text{NVF} = \text{NORM}\left\{\frac{1}{1+\delta^2_{\text{block}}}\right\} \tag{1.6}$$

where δ is luminance variance of the computed blocks.

High values of PSNR and WPSNR indicate more imperceptibility. The robustness of any image watermarking scheme is high if the NC and SSIM values are close to 1 (Thanki et al., 2017).

The NC and SSIM are calculated using Equations 1.7 and 1.8, respectively. While NC measures the correlation between original and extracted watermarks, SSIM measures the structural similarity between them.

$$\text{NC} = \frac{\sum_{x=1}^{M}\sum_{y=1}^{N} w(x,y) \times w*(x,y)}{\sum_{x=1}^{M}\sum_{y=1}^{N} w^2(x,y)} \tag{1.7}$$

$$\text{SSIM} = \frac{\left(2\mu_w \mu_{w^*} + C_1\right)\left(2\sigma_{ww^*} + C_2\right)}{\left(\mu_w^2 + \mu_{w^*}^2 + C_1\right)\left(\sigma_w^2 + \sigma_{w^*}^2 + C_2\right)} \tag{1.8}$$

where C_1 and C_2 are constants, w and w^* are the original and extracted watermarks, μ_w and μ_{w^*} are the means of the original and extracted watermarks, σ_{ww} and σ_{ww^*} represent the covariance of the original and extracted watermarks, and N represents the number of windows.

1.7 ATTACKS ON WATERMARKS

Although there are plenty of watermarking algorithms and tools available online, watermarking is still a challenging task when dealing with attacks (Wolfgang et al., 1999). An attack, in watermarking terminology, aims to remove traces of the watermark to thwart authentication goals. Knowing the attack types and their analysis assists in identifying the best tool for a particular application in competing environments. Checkmark, Optimark, and StirMark are some of the software that assess the robustness of developed watermarking tools (Borra et al., 2017). The attacks are generally categorized into removal, geometrical, cryptographic, oracle, protocol, and security attacks.

- Removal attacks aim to remove watermarks from a watermarked image by intentionally performing image processing operations, such as quantization, lossy compression, averaging, remodulation, demodulation, collusion attacks, block replacement attacks, denoising, and filtering. It has been observed that to perform a remodulation attack, the attacker first has to forecast the watermark using a variety of filters (median, high pass, and Wiener), then subtract it from the watermarked image, and finally add the Gaussian noise to it. This is commonly referred to as a collusion attack. However, as a mosaic attack, the attacker splits the watermarked image into small portions and tries to reassemble it using an HTML table, with the intention of removing the inserted watermark.
- Geometrical attacks aim to distort the watermark rather than remove it by creating nonsynchronization among the extracted and original watermarks. The attackers try to crop the image from its sides or delete/edit/shift some rows or columns of pixels randomly with the intention of distorting the watermark. This creates synchronization problems, while extracting the watermarks. It is common to apply a combination of attacks rather than a single attack. The common geometrical attacks are rotation, shearing, translation, affine transformation, scaling, aspect ratio changes, cropping, column/line removals, jitter, and random bending. Note that

these attacks can be applied either locally or globally and in a systematic or random way.

- Protocol attacks aim to create ambiguity of ownership by attacking the watermarking application itself, using the concept of invertible watermarks. Protocol attacks set another requirement for the design of watermarking tools: watermark extraction must be impossible from any images that are in fact not watermarked. Said another way, the false-positive rate must be at a minimum. Also, protocol attacks take advantage of loopholes in the management or implementation process of watermarking.
- Copy attacks aim to predict a watermark and replicate it on other data without knowledge of the secret keys involved. Also, development of image-dependent watermarks is the best solution to thwart these kinds of attacks.
- Legal attacks attempt to create doubt in the technical evidence on watermarks and watermarking schemes, while proving ownership in the courts.
- Cryptographic attacks envisage finding the lengthy secret keys by exhaustive searches.
- Oracle attacks aim to generate the original host image from the watermarked image using a watermark detector algorithm.
- Disable detection attacks aim to break the relationship between the watermark and host, which carries it without affecting the existence of the watermark.
- Ambiguity attacks embed multiple fake watermarks in order to mislead the detector. Ambiguity attacks occur in systems with multiple watermarks. In such cases, sometimes the order in which the watermarks are inserted is ambiguous.

1.8 DIGITAL IMAGE WATERMARKING TOOLS

Piracy has been a serious concern, as illegally distributed copies are leading to a huge loss to IP owners due to the rapid uploading of multimedia files over the Internet. Watermarking tools are employed as a way to protect IP rights and prevent illegal forgery and piracy. A variety of image watermarking tools are available online for watermark creation, embedding, and extraction (Borra et al., 2017). While some tools restrict the watermarks to only those available in their database, other tools allow the creation of one's own watermark. Table 1.1 lists the web links of 26 image watermarking tools and 3 robustness verification tools presently available online.

After analyzing the case studies of customers using the aforementioned watermarking tools, it was noticed that most professional artists/photographers still employ visible watermarks to protect their artworks/photos at multiple scales, as the most challenging task of resolving the trade-off between robustness and invisibility is unsolved.

With invisible watermarks, their purpose is served only if they robustly survive in the host image irrespective of a variety of defined and undefined attacks. The development of intelligent digital image watermarking techniques based on advanced machine learning, neural networks, artificial intelligence, and bio-inspired algorithms in conjunction with a variety of transforms is required for better trade-off among several requirements of watermarking. The algorithms for finding optimal embedding parameters (loca tions/intensities/coefficients/subbands) based on image characteristics are the focus of research. Reducing the computational time and complexity of such intelligent algorithms and dealing with issues related to the local optimum are also challenging research problems. The extended requirements of watermarking tools may include real-time processing, mathematical formulations and support to multiple resolutions, multiple formats, and multimedia. Furthermore, automated techniques and benchmarks for the proper and simple assessment of watermarking tools for a particular application need to be developed.

TABLE 1.1 Digital Image Watermarking Tools

Sl. No.	Tool Name	Description
01	Umark Lite	https://www.uconomix.com/
02	WTM	http://www.pearlmountainsoft.com/watermark/
03	WMT PLUS	https://www.pearlmountainsoft.com/watermarkplus/
04	Water marquee	https://www.watermarquee.com/
05	Lunapic	http://www196.lunapic.com/editor/
06	Visual water mark	https://www.visualwatermark.com/
07	Mass water mark	http://www.masswatermark.com/
08	Water mark lib	http://download.cnet.com/WatermarkLib/3000-2192_4-10963794.html
09	Alamoon	http://alamoon.com/
10	TSR water mark	https://www.watermark-image.com/
11	1-More water marker	https://1-more-watermarker.en.softonic.com/
12	Photo Watermark Professional	https://archive.org/details/tucows_256118_Photo_Watermark_Professional
13	Fast water mark	https://fast-watermark.en.softonic.com/
14	Watermark Master	http://www.videocharge.com/Products/wm/main.php
15	Watermark.ws	https://www.watermark.ws/
16	Cooltweak	http://www.cooltweak.com/
17	Picture stamper	http://amin-ahmadi.com/picture-stamper/
18	Water mark passion	http://www.majorgeeks.com/mg/getmirror/watermark_passion,1.html

(*Continued*)

TABLE 1.1 (CONTINUED) Digital Image Watermarking Tools

Sl. No.	Tool Name	Description
19	Easy watermark studio lite	http://www.easy-watermark-studio.com/easy-watermark-studio/lite-version.html
20	Bytescout watermarking	https://bytescout.com/products/enduser/watermarking/watermarking.html
21	Snagit	http://snagit1.software.informer.com/download-snagit-watermarks/
22	Jaco Watermark	http://jaco-watermark.sourceforge.net/
23	Star Watermark	http://www.star-watermark.com/
24	Arclab Watermark	https://www.arclab.com/en/watermarkstudio/
25	Batch Watermark	http://batch-watermark-creator.software.informer.com/6.0/
26	Optimark	http://poseidon.csd.auth.gr/optimark/
27	StirMark	http://www.petitcolas.net/fabien/watermarking/stirmark/index.html

REFERENCES

Abdelhakim, Assem M., M. H. Saad, M. Sayed, and H. I. Saleh. "Optimized SVD-based robust watermarking in the fractional Fourier domain." *Multimedia Tools and Applications* 1–23, 2018.

Biswas, Debalina, Poulami Das, Prosenjit Maji, Nilanjan Dey, and Sheli Sinha Chaudhuri. "Visible watermarking within the region of non-interest of medical images based on fuzzy C-means and Harris corner detection." *Computer Science & Information Technology* 24: 161–168, 2013.

Borda, Monica E. "Digital rights protection—a great challenge of the new millennium." In *7th International Conference on Telecommunications in Modern Satellite, Cable and Broadcasting Services*, vol. 1, pp. 207–214. IEEE, Piscataway, NJ, 2005.

Borra, Surekha, Rohit Thanki, Nilanjan Dey, and Komal Borisagar. "Secure transmission and integrity verification of color radiological images using fast discrete curvelet transform and compressive sensing." *Smart Health* 2018.

Borra, Surekha, H. Lakshmi, N. Dey, A. Ashour, and F. Shi. "Digital image watermarking tools: state-of-the-art." In *Information Technology and Intelligent Transportation Systems: Proceedings of the 2nd International Conference on Information Technology and Intelligent Transportation Systems*, Xi'an, China, vol. 296, p. 450, 2017.

Dey, Nilanjan, Sayan Chakraborty, and Sourav Samanta. *Optimization of Watermarking in Biomedical Signal*. Lambert Academic Publishing, Riga, Latvia, 2014.

Dey, Nilanjan, Amira S. Ashour, Fuqian Shi, Simon James Fong, and R. Simon Sherratt. "Developing residential wireless sensor networks for ECG healthcare monitoring." *IEEE Transactions on Consumer Electronics* 63(4): 442–449, 2017.

Dey, Nilanjan, Amira S. Ashour, Sayan Chakraborty, Sukanya Banerjee, Evgeniya Gospodinova, Mitko Gospodinov, and Aboul Ella Hassanien. "Watermarking in biomedical signal processing." In *Intelligent Techniques in Signal Processing for Multimedia Security*, pp. 345–369. Springer, Cham, 2017.

Dey, Nilanjan, and V. Santhi, eds. *Intelligent Techniques in Signal Processing for Multimedia Security*. Springer, Cham, 2017.

Hartung, Frank, and Martin Kutter. "Multimedia watermarking techniques." *Proceedings of the IEEE* 87(7): 1079–1107, 1999.

Kutter, Martin, and Fabien A. P. Petitcolas. "Fair benchmark for image watermarking systems." In *Security and Watermarking of Multimedia Contents*, vol. 3657, pp. 226–240. International Society for Optics and Photonics, Bellingham, WA, 1999.

Langelaar, Gerhard C., Iwan Setyawan, and Reginald L. Lagendijk. "Watermarking digital image and video data. A state-of-the-art overview." *IEEE Signal Processing Magazine* 17(5): 20–46, 2000.

Mohanty, Saraju P., Anirban Sengupta, Parthasarathy Guturu, and Elias Kougianos. "Everything you want to know about watermarking: from paper marks to hardware protection." *IEEE Consumer Electronics Magazine* 6(3): 83–91, 2017.

Petrovic, Rade, Babak Tehranchi, and Joseph M. Winograd. "Digital watermarking security considerations." In *Proceedings of the 8th Workshop on Multimedia and Security*, pp. 152–157. ACM, New York, 2006.

Qasim, Asaad F., Farid Meziane, and Rob Aspin. "Digital watermarking: applicability for developing trust in medical imaging workflows state of the art review." *Computer Science Review* 27: 45–60, 2018.

Sherekar, S. S., V. M. Thakare, and Sanjeev Jain. "Critical review of perceptual models for data authentication." In *International Conference on Emerging Trends in Engineering and Technology*, pp. 323–329. IEEE, Piscataway, NJ, 1999.

Surekha, B., and G. N. Swamy. "Security analysis of 'A novel copyright protection scheme using visual cryptography.'" In *International Conference on Computing and Communication Technologies (ICCCT)*, pp. 1–5. Hyderabad, India, 2014.

Thanki, Rohit M., and Ashish M. Kothari. "Digital watermarking: Technical art of hiding a message." In *Intelligent Analysis of Multimedia Information*, pp. 431–466. IGI Global, Hershey, PA, 2017.

Thanki, Rohit, Surekha Borra, Nilanjan Dey, and Amira S. Ashour. "Medical imaging and its objective quality assessment: an introduction." In *Classification in BioApps*, pp. 3–32. Springer, Cham, 2018.

Thanki, Rohit M., and Ashish M. Kothari. "Digital watermarking: Technical art of hiding a message." In *Intelligent Analysis of Multimedia Information*, pp. 431-466. IGI Global, Hershey, PA, 2017.

Thanki, Rohit M., Surekha Borra, Vedvyas Dwivedi, and Komal Borisagar. "An efficient medical image watermarking scheme based on FDCuT-DCT." *Engineering Science and Technology, an International Journal* 20(4): 1366–1379, 2017.

Vellasques, E., E. Granger, and R. Sabourin. "Intelligent watermarking systems: A survey." In *Handbook of Pattern Recognition and Computer Vision*," pp. 687–724. World Scientific Publishing, Singapore, 2010.

Wang, Zhou, and Alan C. Bovik. "A universal image quality index." IEEE *Signal Processing Letters* 9(3): 81-84. IGI Global, Hershey, PA, 2002.

Wolfgang, Raymond B., Christine I. Podilchuk, and Edward J. Delp. "Perceptual watermarks for digital images and video." *Proceedings of the IEEE* 87(7): 1108–1126, 1999.

CHAPTER 2

Advanced Watermarking Techniques

When transmitting images over an open communication channel, different threats are encountered related to confidentiality, authenticity, and integrity. Different kinds of digital watermarking approaches can be considered to warn, prevent, limit, and control access, apart from proving ownership. Recent developments of digital media, which has replaced paper, are facing issues related to piracy, security, and copyrights. Digital watermarking techniques are developed with the intention of protecting multimedia over such threats, by integrating a watermark either visibly or invisibly in a robust manner before publishing or storing the media. Images are often impaired unintentionally while storing, transmitting, retrieving, converting, and viewing. When designing robust watermarking systems, the watermarks are preferably embedded in the most perceptually significant parts of an image to ensure its robustness even to intentional distortions, such as translation, scaling, rotation, and affine transformations. Being able to resist such attacks, apart from general noise, while maintaining transparency and security is required for copy control, transaction tracking, and proof of image ownership. For three decades researchers and developers have been developing algorithms, architectures, and systems in this area to gain the best results.

This chapter discusses various conventional and machine learning (ML)–based digital image watermarking algorithms and compares some recent works to define the challenges to be considered for future research. The chapter starts with spatial domain–based watermarking approaches, followed by transform domain techniques, and ends with advanced watermarking techniques that rely on the concept of ML algorithms.

2.1 INTRODUCTION

In today's e-era, which mainly includes e-commerce, e-banking, e-health, and e-learning, information can be effectively downloaded with no authorization from the proprietor. Once in a while, these circumstances create issues, for example, copyright security and proprietor verification. In such cases, assurance and verification of advanced information is required before it is exchanged over an open-source transmission medium. To address these issues, analysts have proposed different information-concealing strategies: cryptography, steganography, and watermarking. Watermarking is primarily utilized for the assurance and validation of information. This method beats the constraints of steganography by embedding a watermark into host content such that even the basic client cannot find the hidden watermark (Borra and Lakshmi, 2015). As discussed in Chapter 1, the watermarking framework has three components: a watermark embedder; a correspondence channel, which might be wired or wireless; and a watermark extractor. The watermark embedder embeds a watermark into host images to generate a watermarked image, while the watermark extractor extracts the watermark from the test image, which can be the watermarked image with or without attacks. The major requirements of digital image watermarking are recalled here:

1. Robustness: The watermarking technique must protect owners' data against any manipulations and has to be robust.
2. Imperceptibility: After insertion of the watermark into host data, the visual quality of the host data should not be affected much; that is, the watermark should not be perceptible.
3. Embedding capacity: The watermark technique should allow hiding of large watermarks (Banerjee et al., 2015).

The watermarking techniques are majorly developed in two processing domains: the spatial domain and transform domain (Borra and Lakshmi, 2015). The spatial domain techniques are easy to implement but provide less imperceptibility as the host image pixels are directly modified. Transform domain watermarking is complex but provides more robustness than spatial domain watermarking. In all the transform domain techniques, the host image is converted into the frequency domain using various image transforms, such as discrete Fourier transform (DFT), discrete cosine transform (DCT), and discrete wavelet transform (DWT) (Dey et al., 2012), before watermark embedding, and later the image is inverse transformed. The framework for conventional digital image watermarking is shown in Figure 2.1. In spatial domain watermarking schemes, the host image is modified directly by changing the least significant bits (LSBs) or adding various noise sequences. In transform domain watermarking, the watermarks are embedded into either selected domain coefficients or blocks (Hartung and Kutter, 1999; Bender et al., 1996; Surekha and Swamy, 2011; Ashour and Dey, 2017; Thanki and Kothari, 2017).

The preprocessing of host images before watermark embedding helps us extract better features for watermark embedding

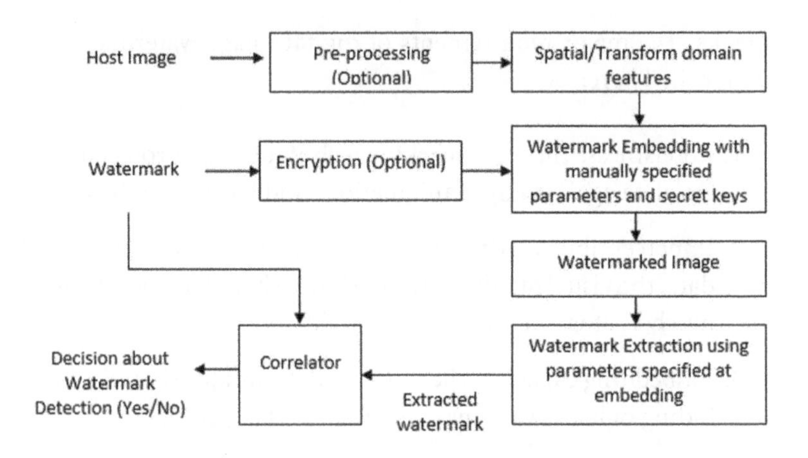

FIGURE 2.1 Framework of conventional digital image watermarking.

and results in better performance. Normalization, segmentation, and feature extraction are common preprocessing techniques. Normalization of an image leads to robustness to illumination changes and scale variances. Segmentation and feature extraction result in achieving high robustness to intentional attacks, such as geometric variations. Techniques such as Gaussian mixture model (GMM), expectation maximization (EM), mean-shift segmentation, Mexican hat wavelet scale interaction, scale-invariant feature transform (SIFT), and difference of Gaussians (DoG), can be employed to resist attacks, which include rotation, translation, scaling, low-pass filtering, median filtering, Gaussian noise, aspect ratio changes, and JPEG compression. The detection and/or verification of a watermark refers to the process of making a binary decision (presence/absence of a particular watermark) at the correlator about the authentication of the rightful owner.

2.2 WATERMARKING IN THE SPATIAL DOMAIN

Spatial domain techniques support embedding high-capacity watermarks directly into the pixels of the host image with low complexity in less time and in a highly controllable manner

(Surekha and Swamy, 2012a,b). This section discusses a variety of spatial domain watermarking techniques.

2.2.1 Least-Significant-Bit Substitution Technique

The LSB substitution technique is a basic watermarking technique where less important information or bits of host data are modified by the watermark. In the LSB technique, most significant bits of the watermark are substituted in the LSBs of the host data (Lee and Chen, 2000; Chan and Cheng, 2004; Ramalingam, 2011). In any 8-bit image, the most significant bit plane (i.e., bit plane 7) contains the most important visual information, while the last or least significant bit plane (i.e., bit plane 0) contains no visual information. All other bit planes contribute to various levels of information related to the image. Therefore, the least significant bit plane of the image is chosen for watermarking purposes. For example, to embed an 8-bit gray watermark into a color image, the gray watermark bits are divided into groups of 3 bits, 3 bits, and 2 bits. The first two group's bits can be inserted into the last three LSB bits of the R channel and G channel. The group of 2 bits can be inserted into the two LSB bits of the B channel. Then, these three channels are combined to generate a watermarked color image. The main advantage of this technique is that after the watermark embedding process, the visual quality of the host data is not much affected by the watermark. Hence, under normal conditions, an average individual cannot see or observe the modifications in the host image where the watermark is inserted. The payload capacity of this technique is almost 100%. This technique is mainly used for copyright authentication of multimedia data. The limitation of this technique is that watermarks embedded in this way are fragile in nature.

2.2.2 Patchwork Technique

Patchwork (Bender et al., 1996) uses some statistics to generate random patterns, which are embedded into the host image (Durvey and Stayarthi, 2014; Soman, 2010) to provide robustness against lossy compression and various image processing manipulations, such as filtering and the addition of noise. The technique can embed only small watermarks into the host image. Embedding of large watermarks into host images is by block processing. For every bit of watermark, a pair of blocks is considered. To hide bit 1, the intensity of one block is increased and the intensity of another block in the pair is decreased by a fixed factor α. No blocks are altered if the watermark bit is 0. The positions of pairs of blocks depend on the secret key. To detect the watermark, the mean sum of differences between the pair of blocks is calculated and compared with a threshold. The location of the blocks is kept secret, and assuming certain properties for image data, the watermark is easily located by averaging the difference between the values in the two subsets. It is assumed that, on average, without the watermark, this value goes down to zero for image data.

2.2.3 Texture Mapping Coding Technique

This technique is used for the copyright protection of images that are rich in texture information (Datta and Nath, 2014). A host image is first segmented into two regions: texture and nontexture regions. Later, the pixels in the texture region are modified by a watermark using any of the spatial domain watermark embedding procedures, followed by combination with a nontexture region of the host image to obtain the watermarked image. In the extraction process, first the watermarked image is divided into a nontexture region and texture region. Then, by reversing the procedure, the watermark is recovered from the watermark texture region. This technique is mainly used for the copyright protection of images related to medical science and defense.

2.2.4 Predictive Coding Technique

This technique was designed and implemented by Matsui and Tanaka for grayscale images (Matsui, 1998). In this technique, the correlation between adjacent pixels is calculated. A watermark is generated by a set of pixels that are embedded, and every other pixel is modified by the differences between adjacent pixels. These techniques are more robust than the LSB substitution technique (Pandhwal and Chaudhari, 2013; Saxena et al., 2015).

2.2.5 Additive Watermarking Technique

The traditional method of watermarking is to employ M sequences/pseudorandom noise (PN) sequences to sequentially or randomly modify the blocks of host images. Usage of PN sequences allows blind detection of watermarks due to their excellent correlation properties. These approaches exploit usage of PN sequences in embedding the scaled watermarks in an additive fashion to the host image (Langelaar et al., 2000; Arena et al., 2000; Bangaleea and Rughooputh, 2002; Shoemaker, 2002). These sequences are preferred as they can affect the pixel values in an imperceptible way due to their low magnitudes. The random sequences are generated using a seed when required, which acts as a secret key. Two noise sequences are usually generated using the secret key. One sequence is used for insertion of watermark bit 0, and the other is used for watermark bit 1. The watermark is inserted into the host image using the following steps:

1. Take the host image and divide it into blocks. The size of the blocks depends on the size of watermark and host image.
2. Generate two highly uncorrelated noise sequences using the secret key.
3. Create a watermark mask using these noise sequences and the watermark as follows:
    - If the watermark bit is zero, then fill the mask with one type of noise sequence; otherwise, fill the mask with another noise sequence.
    - Repeat this procedure for all blocks of the host image.

4. Add the generated watermark mask to the host image with the help of a gain factor using the additive watermark equation below.

$$\mathrm{WI}(x,y) = I(x,y) + k \times W(x,y) \tag{2.1}$$

where $\mathrm{WI}(x, y)$ represents the watermarked image, $I(x, y)$ represents the host image, k represents the gain factor, and $W(x, y)$ represents the watermark mask.

The additive watermarks are extracted from the watermarked data using the following procedure:

1. Divide the watermarked data into blocks, keeping the size of the blocks the same as that of the PN sequence.
2. Calculate the correlation between each block of watermarked data and each of the noise sequences.
3. Extract the watermark using the following two conditions:
   - If the correlation result corresponding to noise sequence 1 is higher than the correlation result corresponding to noise sequence 0, then set the watermark to 1.
   - Otherwise, set the watermark bit to 0.
4. Repeat the procedure for all blocks of watermarked image to extract the watermark.

2.2.6 Other Spatial Domain Watermarking Techniques

In additive watermarking techniques, the watermark blocks are selected sequentially to hide watermark bits, and hence are not robust to many attacks. In spread-spectrum-based watermarking techniques, the watermark bits are randomly scattered and can even be repeatedly embedded throughout the host image in order to increase the security and robustness against attacks. Local binary pattern (LBP)–based watermarking is good at surviving illumination changes and contrast variations. The host image is divided into nonoverlapping neighborhoods. Every pixel in the neighborhood is then compared with the center pixel, so as to produce a binary valued matrix, which is further used for watermark embedding. Since LBP-based methods can survive few attacks, their application is limited to fragile watermarking. Watermarks can even be hidden by modifying or shifting the local or global histograms of an image without affecting their shape. However, these methods support small watermarks.

For color host images, the blue component can be considered for watermarking to maintain imperceptibility. Other novel approaches for watermarking image data include fractal-based approaches (Essaouabi and Ibnelhaj, 2009; Raghavendra and Chetan, 2009) and geometric feature–based watermarking (Hussein and Mohammed, 2009). Here, salient points in an image are found and warped according to a dense line pattern representing the watermark and generated randomly. The detection process involves determining whether a significantly large number of points are within the vicinity of the line patterns.

2.3 WATERMARKING IN THE TRANSFORM DOMAIN

DFT converts any digital function into its frequency coefficients. In digital signal or image processing, the function that is aperiodic in nature can be represented by an integral of sine and/or cosine values. These values make Fourier coefficients of the function. This transform is robust against various geometric operations, such as rotation, scaling, and translation (RST) and cropping. DFT allows the analysis and processing of any function in the frequency domain using its Fourier coefficients. There are many watermarking approaches based on DFT (Ramkumar et al., 1999; Deguillaume et al., 1999; Lin et al, 2001; Solachidis and Pitas, 2001; Solachidis and Pitas, 2004; Potdar et al., 2005; Hendriks et al., 2008; Kaushik, 2012; Ansari et al., 2012). The simple approach based on DFT involves the application of DFT on the host image, the coefficients of which are modified according to the watermark or encrypted watermark with a gain factor and private key using the additive watermarking approach. After modified DFT coefficients are obtained, inverse DFT is applied to obtain the watermarked image. In the extraction process, the inverse process of watermark embedding is followed to recover the watermark from the watermarked image. This approach can be designed to be non-blind and can use encryption techniques, such as Arnold scrambling, for securing the watermark.

An alternative approach to using DFT is to divide the host image into nonoverlapping blocks. Then DFT is applied on blocks of the host image to obtain DFT coefficients, which are then modified according to PN sequences and watermark bits. The inverse DFT is applied on modified blocks of the host image to obtain the watermarked image. At extraction, correlation between PN sequences and DFT coefficients of blocks of the watermarked image is performed to extract the watermark. This is a blind approach and provides better robustness than the first approach. Many watermarking approaches that combine DFT with other transforms, such as DCT and DWT, have been developed to improve the imperceptibility and robustness of watermarking (Ansari et al., 2012). DFT-based watermarking is robust to geometric attacks such as RST. The drawback is that the data is represented in a complex form and hence provides fewer transiencies for watermarked data.

2.4 WATERMARKING IN THE DISCRETE COSINE TRANSFORM

DCT is a linear transform that maps an $n \times n$ size image into another $n \times n$ size matrix with elements representing the distribution and energy of frequency components of the image. DCT (Langelaar et al., 2000; Arena et al., 2000; Hernandez et al., 2000; Lu and Liao, 2001; Shoemaker, 2002; Huang and Guan, 2004; Preda and Vizireanu, 2007; Sridevi et al., 2010; Ding et al., 2010) converts the host image into its cosine coefficients, which represents DC as well as AC coefficients. The DC coefficient is located at the upper left corner of the matrix and contains very significant information. The DC coefficient is an integer and represents the average color of the image. The AC coefficients may or may not be integers. Figure 2.2 shows the DCT coefficient distribution of blocks of an image.

The significant information in an image is located in the low-frequency DCT coefficients. The high-frequency DCT coefficient, which represents the sharp details in an image, can be easily removed by blurring and Gaussian noise attacks. Therefore, for robust watermarking, midfrequency DCT coefficients are used. There are two main approaches available in the literature for watermark embedding in the DCT domain: (1) mid-band-frequency DCT coefficient comparison (Langelaar et al., 2000; Shoemaker, 2002) and (2) addition of PN noise sequences to mid-band-frequency DCT coefficients (Langelaar et al., 2000; Shoemaker, 2002). Both approaches are blind, do not require the original host image for watermark extraction, and are widely used for copyright protection of multimedia data (Langelaar et al., 2000; Shoemaker, 2002). An alternative approach is to consider a set of large DCT coefficients for modification. In the non-blind approach, DCT

LF	LF	LF	MF	MF	MF	MF	HF
LF	LF	MF	MF	MF	MF	HF	HF
LF	MF	MF	MF	MF	HF	HF	HF
MF	MF	MF	MF	HF	HF	HF	HF
MF	MF	MF	HF	HF	HF	HF	HF
MF	MF	HF	HF	HF	HF	HF	HF
MF	HF	HF	HF	HF	HF	HF	HF
HF	HF	HF	HF	HF	HF	HF	HF

FIGURE 2.2 Frequency distribution in DCT. LF = low-frequency coefficients; MF = middle-frequency coefficients; HF = high-frequency coefficients.

coefficients of the host image are modified according to the following steps to generate a watermarked image:

1. Calculate the size of the host image.
2. Apply forward DCT to the host image to get its DCT coefficients, such that

$$D = \mathrm{DCT}(I) \tag{2.2}$$

   where D represents the DCT coefficients of the host data.
3. Calculate the size of the watermark.
4. Insert the watermark into the DCT coefficients of the host data using the equation below.

$$D'_{x,y} = D_{x,y} + k \times w_{x,y} \tag{2.3}$$

   where D' is modified DCT coefficients, w is the watermark, and k is the gain/scaling factor.
5. Apply inverse DCT on modified DCT coefficients to get the watermarked data.

$$\mathrm{WI} = \mathrm{IDCT}(D') \tag{2.4}$$

   where WI is the watermarked image.

The steps of the basic watermark extraction process in the DCT domain are given below.

1. Calculate the size of the watermarked image.
2. Apply forward DCT on the watermarked image to get its DCT coefficients.

$$D' = \mathrm{DCT}(\mathrm{WI}) \tag{2.5}$$

   where D' represents the DCT coefficients of the watermarked image.

3. Extract the watermark from the DCT coefficients of the watermarked image using the below equation.

$$w'_{x,y} = (D'_{x,y} - D_{x,y}) / k \tag{2.6}$$

where w' is an extracted watermark.

Watermarks embedded in midfrequency coefficients of DCT are robust to common attacks but can support much smaller payloads. While watermarks embedded in high frequency are not robust to filtering/blurring attacks, they may be preferred for fragile watermarking. DCT is robust to JPEG compression. DCT-based techniques show poor robustness to affine and geometric attacks, such as shifting, cropping, rotation, and shearing. A few hybrid watermarking approaches based on the combination of DCT and other transforms are available in the literature (Huang and Guan, 2004). These approaches are mainly designed to improve imperceptibility and robustness, which cannot be achieved using a single DCT-based approach.

2.5 WATERMARKING IN THE DISCRETE WAVELET TRANSFORM

In many image processing applications, such as data compression, watermarking, and data fusion, DWT plays a very important role due to its multiresolution capabilities and spatial resolution characteristics. DWT is easy to implement and fast in computation. The image is converted into four subbands using DWT decomposition: approximation subband (LL), horizontal subband (HL), vertical subband (LH), and diagonal subbands (HH). Figure 2.3 shows the basic wavelet decomposition steps.

In DWT-based watermarking methods (Ejima and Miyazaki, 2000; Serdean et al., 2002; Raval and Rege, 2003; Fan and Yanmei, 2006; Elbasi, 2007; Essaouabi and Ibnelhaj, 2009; Raghavendra and Chetan, 2009; Hussein and Mohammed, 2009; Mostafa et al., 2009), coefficients of the wavelet subband of the host image are modified according to the watermark and gain

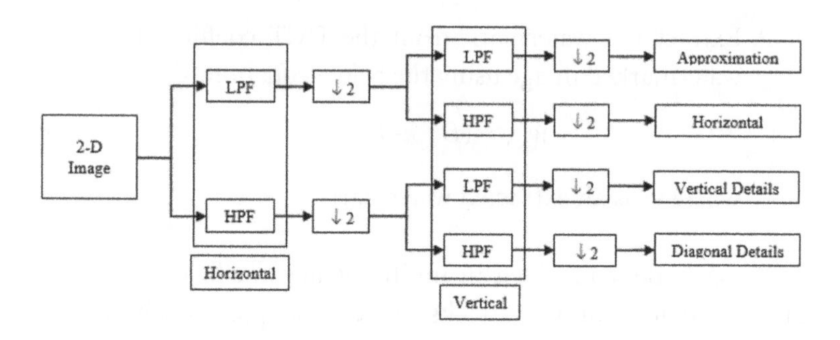

FIGURE 2.3 Basic wavelet decomposition of image.

factor. In the robust and invisible watermarking algorithm, first the host image is divided into nonoverlapping blocks, followed by application of DWT to obtain wavelet coefficients of blocks of the host image. One or more wavelet subband coefficients are then chosen to be modified for watermark embedding using approaches very similar to DCT. The inverse DWT is applied on modified coefficients with other unmodified coefficients to obtain the watermarked image. On the extraction side, the watermark can be extracted either blindly or in a non-blind manner based on the type of embedding, using secret keys and scaling factors.

The most significant information of an image is located in the approximation subband, and less significant information is distributed into other wavelet subbands. DWT-based methods survive against low-pass filtering and median filtering but are not robust to geometric attacks. Although there are many DWT-based hybrid approaches, the combination of DWT with SVD is widely used and most popular in the research community. For fragile watermarking, the LL subband of a host image can be chosen for embedding, but the drawback is poor imperceptibility.

2.6 WATERMARKING USING SINGULAR VALUE DECOMPOSITION

Singular value decomposition (SVD) (Thakkar and Srivastava, 2017; Su et al., 2013; Kamlakar et al., 2012; Gupta and Raval, 2012;

Mansouri et al., 2009; Santhi and Thangavelu, 2009; Rajab et al., 2008; Dili and Mwangi, 2007; Ganic and Eskicioglu, 2004) is a numerical technique based on linear algebra and is used to diagonalize matrices in numerical analysis. There are several areas where SVD finds application. The application of SVD to digital data of size $M \times N$ results in three matrices: U, V, and S. The U and V matrices are called unitary matrices and have size $M \times N$. The S matrix is called a singular or diagonal matrix and has size $M \times N$. The singular matrix plays an important role in watermarking; entries in this matrix are arranged diagonally and in ascending order. The singular values are very stable; hence, if a small change is made in the value of host data, its singular values do not undergo any significant change.

Various types of SVD-based watermarking methods are available in the literature. In these methods, selected singular values of the host image are modified by watermarks using a gain factor. In direct modification of the singular values method, SVD is applied on nonoverlapping blocks of the host image to obtain a set of singular values that can further be modified using methods similar to DCT/DWT watermarking. This method is widely used by researchers as it allows embedding of any type of watermark, such as binary, grayscale, and color. This method shows robustness to many attacks and, when used with DWT, provides higher transparency.

Another distinct approach is to make use of the second and third row values of the first column of the U matrix, which are close in value, for watermark embedding (Thakkar and Srivastava, 2017; Su et al., 2013). In this approach, the two values of the U matrix are modified according to a gain factor and watermark bits to blindly extract the watermark by just comparing them. If the comparison results of these two values are higher than the predefined threshold, then the watermark bit is set to 1; otherwise, it is set to 0. This method is robust to many attacks and is a blind approach. The limitation of this method is that it can embed only binary watermarks. The drawback of SVD-based methods is that they are applicable only to symmetric data or high-contrast host images.

2.7 COMPRESSIVE SENSING AND QR DECOMPOSITION METHODS

Around 2005, researchers introduced a new signal processing theory based on the sparsity property of a signal. This theory is known as compressive sensing (CS) (Donoho, 2006; Candes, 2006; Baraniuk, 2007) and its application to watermarking was introduced by Sheikh and Baraniuk in 2007. This theory was used to provide security by adding two processes—CS-based encryption and CS-based decryption—to the conventional watermarking approaches. The necessary condition for application of the CS theory to an image is that the image must be sparse in its own domain. Thus, first the watermark is converted into its sparse coefficients using various image transforms, such as DFT, DCT, DWT, and SVD, for the generation of sparse coefficients of the watermark. The drawback with encrypting a watermark before its embedding is that the computational time is increased. This technique is mainly used for copyright authentication and tamper detection using identification of multimedia data (Sheikh and Baraniuk, 2007; Valenzise et al., 2009; Zhang et al., 2011; Raval et al., 2011; Fakhr, 2012; Tiesheng et al., 2013; Thanki et al., 2017, 2018). The CS theory can even be used to encrypt a host image before the watermark is embedded into it.

QR decomposition decomposes a matrix into an orthogonal matrix and a triangular matrix. QR decomposition of a real square matrix A is represented as $A = QR$, where Q is an orthogonal matrix (i.e., $Q^T Q = I$) and R is an upper triangular matrix. If A is nonsingular, then this decomposition is unique. There are two important characteristics of a matrix obtained by QR decomposition (Wang et al., 2016):

1. The value of the first row of elements of the R matrix is larger than that of the elements of another row, indicating that the first row has more energy and hence the R matrix determines the nature of the original image.

2. The value of the first column of elements of the Q matrix represents the relationship between the values of the matrix. Thus, they can provide robustness against common attacks. Therefore, in the QR decomposition–based watermarking technique, a watermark can be inserted into elements of the first column of the Q matrix (Su et al., 2014, 2017; Wang et al., 2016; Rasti et al., 2016; Mehta et al., 2016; Laur et al., 2015; Han et al., 2012; Mitra et al., 2012; Song et al., 2011; Naderahmadian and Hosseini-Khayat, 2010).

2.8 SCHUR DECOMPOSITION–BASED WATERMARKING

Schur decomposition (Van Loan, 1996), when applied on a real matrix A, results in two matrices U and D such that $A = U \times D \times U^T$, where U is a unitary matrix and D is an upper triangular matrix. The D matrix has real eigenvalues on the diagonal. This decomposition requires about $8n^3/3$ computational operations, which is less than those required for SVD decomposition (about $11n^3$). The U matrix has the characteristic that all of the elements of the first two columns have the same sign and values. This property of the U matrix is used in the watermark embedding process. Thus, the blind watermarking approach can be easily designed using this decomposition. Schur decomposition–based watermarking (Li et al., 2017; Rajab et al., 2015; Meenakshi et al., 2014; Razafindradina et al., 2013; Gunjan et al., 2012; Mohammad, 2012; Su et al., 2012; Choudhary et al., 2012; Mohan et al., 2011; Mohan and Swamy, 2010; Seddik et al., 2009) provides high transparency compared with all other watermarking approaches, but it is more complex to implement.

2.9 HESSENBERG MATRIX FACTORIZATION IN WATERMARKING

The Hessenberg decomposition (Van Loan, 1996) is a matrix decomposition method. A matrix B is decomposed into a unitary matrix U and a Hessenberg matrix H such that $U \times H \times U^T = B$, where U^T is the conjugate transpose of the unitary matrix. This decomposition requires about $14n^3/3$ computational operations that are higher than the computational operations required in the Schur decomposition but less than those of SVD decomposition. Hessenberg matrix factorization is used in watermarking to ensure that the randomization process is perfectly invertible (Su and Chen, 2017; Singh et al., 2017; Su, 2016; Bhatnagar and Wu, 2013; Bhatnagar et al., 2010). Watermarking using this decomposition is more complex to implement and is less explored by researchers.

2.10 VISIBLE AND REVERSIBLE WATERMARKING

Visible watermarking is a common watermarking approach that is mainly used for content authentication or ownership identification. In this approach, the watermark is visibly overlaid on some portion of the host medium (Biswas et al., 2013). Two sample visible watermarked images generated using existing techniques (Thanki et al., 2011, 2017) are shown in Figure 2.4. The disadvantage of visible watermarks is that they introduce degradation into the host image.

Reversible watermarking is a special case of watermarking. The original host medium is recovered completely at the receiver side, which is not possible in conventional watermarking. In general, any reversible watermarking is implemented in three steps: watermark embedding, watermark extraction, and host medium recovery (Lakshmi and Surekha, 2016; Chakraborty et al., 2014, 2015; Pal et al., 2013). All these processes require a secret key to achieve security of the watermark. A block diagram of reversible watermarking is shown in Figure 2.5. Note that after watermark extraction, there is one more step, called original host image recovery. The input to this algorithm is the watermarked image, secret key, and original watermark, while the output is the extracted original host.

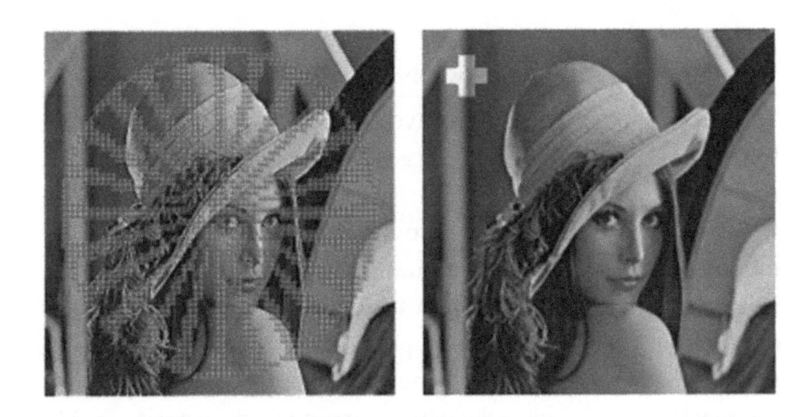

FIGURE 2.4 Sample visible watermarked images.

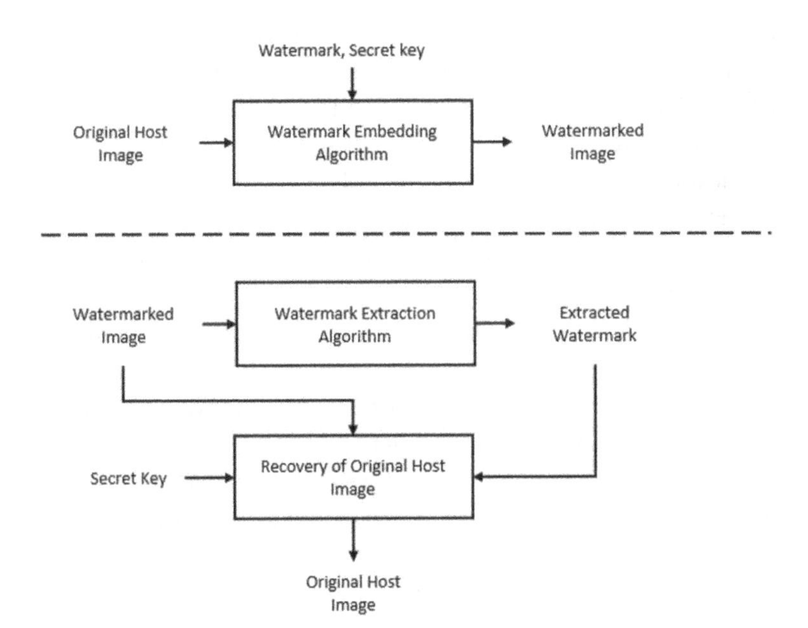

FIGURE 2.5 Block diagram of reversible watermarking.

2.11 MACHINE LEARNING–BASED IMAGE WATERMARKING

The major challenge of conventional image watermarking techniques is to reduce the trade-off among imperceptibility, robustness, capacity, and security. ML, when combined with watermarking, finds optimum solutions for the problem of trade-off minimization. ML techniques automatically predict output values from input data with high accuracy and efficiency by means of different classification and pattern recognition algorithms, which otherwise would be a time-consuming process. ML trains the machine to do the tasks while learning from experience. ML algorithms use various computational and statistical methods to "learn" information directly from data without depending on a predetermined equation or model. The algorithms' performance improves with input samples available for learning. ML algorithms determine natural patterns in input information for better decision making. ML algorithms are used in many real-world applications, such as big data analysis, image processing, computer vision, pattern recognition, object detection in industry, computational biology, natural language processing, and automation in image watermarking.

The models in supervised learning algorithms are based on known input and output information to predict future outputs. The supervised learning algorithms use classification and regression approaches to develop predictive models. While classification approaches are typically used for the predication of discrete data, regression approaches are used for the predication of analog data. The common supervised ML algorithms are decision trees, support vector machines (SVMs), neural networks, nearest neighbor, naïve Bayes, and linear regression. Unsupervised learning algorithms use clustering approaches for the predication of a category of data for analysis. The common ML algorithms under the unsupervised category are *k*-means clustering and association rules.

Artificial neural networks (ANNs) train a system to remember a particular scenario and produce outputs for new inputs based on what it learned. An ANN solves problems based on interconnections of artificial neurons and layers (input, output, and one or multiple hidden layers), into which the learning is divided. Neural network–based watermarking techniques are slow at training and lead to overfitting problems, and their precision is limited to least squares error. Thus, more advanced ML algorithms, such as extreme learning machine (ELM), support vector regression (SVR), and SVM, are employed for intelligent image watermarking. These algorithms overcome the overfitting problems and perform better than gradient-based learning algorithms. SVM and SVRs estimate high-precision nonlinear functions, treating them as convex optimization problems, and hence avoid locking at the local optimal solution. SVM-based techniques work well for small training datasets and high-dimensional data and have proven to be robust against desynchronization attacks.

Most of the ML efforts have recently been directed toward the use of probabilistic neural networks (PNN), which are composed of four layers of nodes: the input, pattern, summation, and output layers. The training of a PNN is performed by generating a pattern node, connecting it to the summation node of the target class, and assigning the input vector as the weight vector. The PNN is a supervised learning network that uses radial basis functions and the Bayes approach for classification. In contrast to back-propagation networks, a PNN works without feedback, but is fast due to parallelism. The fuzzy logic inference system (FIS) closely resembles human reasoning and is employed in digital image watermarking to solve optimization/embedding problems in a flexible way using fuzzifier and defuzzifier with an inference engine and knowledge base, resisting imprecise data.

Recently, researchers introduced application of the convolutional neural network (CNN) algorithm for the security of images using watermarking. The architecture of a CNN is composed of feature detection and classification hidden layers. The feature detection layers perform three types of operations on input data: convolution, pooling, and rectified linear unit (ReLU). The convolution puts the input data through a set of convolutional filters, where each filter works on certain features of input data. The pooling process simplifies the output value by performing nonlinear downsampling. This process reduces the number of output values that the network needs to learn about it. ReLU is used for mapping negative values to zero and only allows positive values. These three operations are repeated over many layers, with each layer learning to extract different features from input data. After feature extraction, classification layers follow where one layer is fully connected and gives an output vector of k dimensions, where k represents the number of input classes that the network predicts. The vector has the probabilities for each input class being classified. The final layer of the CNN architecture uses a classification function to produce a classification output.

All the above-mentioned ML algorithms and others help improve the performance of conventional watermarking techniques by improving the imperceptibility and robustness using four different approaches:

Type I: In this approach, ML algorithms are used to obtain optimized values, with the objective of maintaining the invisibility and quality of the watermarked image by selecting the best positions or intensities or scaling factors for watermark embedding. The generalized block diagram of this approach is shown in Figure 2.6.

Type II: In this approach, an ML algorithm such as an ANN/SVM is modeled as a binary classifier after performing watermark embedding as shown in Figure 2.7. The objective is to memorize the relationship between the watermark and the corresponding watermarked image so as to detect the watermark bits blindly from the test image whenever required, thereby improving the performance of traditional watermarking.

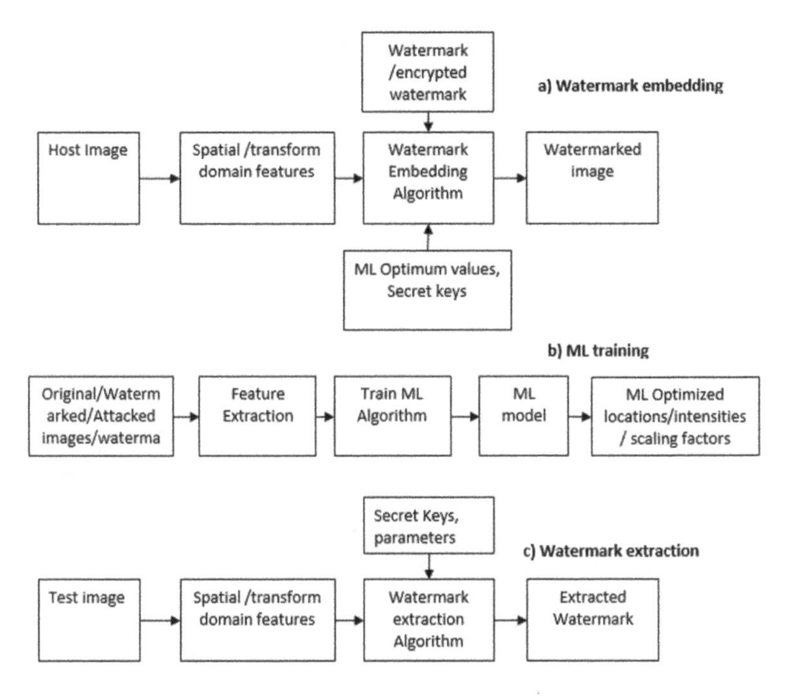

FIGURE 2.6 Type I ML based digital image watermarking.

Type III: In this approach, the ML algorithm is used for watermark embedding and/or extraction algorithms as in Figure 2.8.

Type IV: In this approach, ML algorithms are used for the intelligent enhancement of a watermark's quality after its detection by a conventional extraction algorithm (Tables 2.1 through 2.3).

Recently, deep learning–based watermarking models, whose design flow is similar to that of the CNN, were also introduced. Deep learning is a type of ML approach that learns to perform application tasks directly from input data. A deep learning algorithm combines multiple nonlinear processing layers using simple elements operated in parallel that are inspired by the human nervous system. Each layer is interconnected via nodes or neurons, where each hidden layer generates results based on outputs generated by the previous layer. In contrast to conventional ML

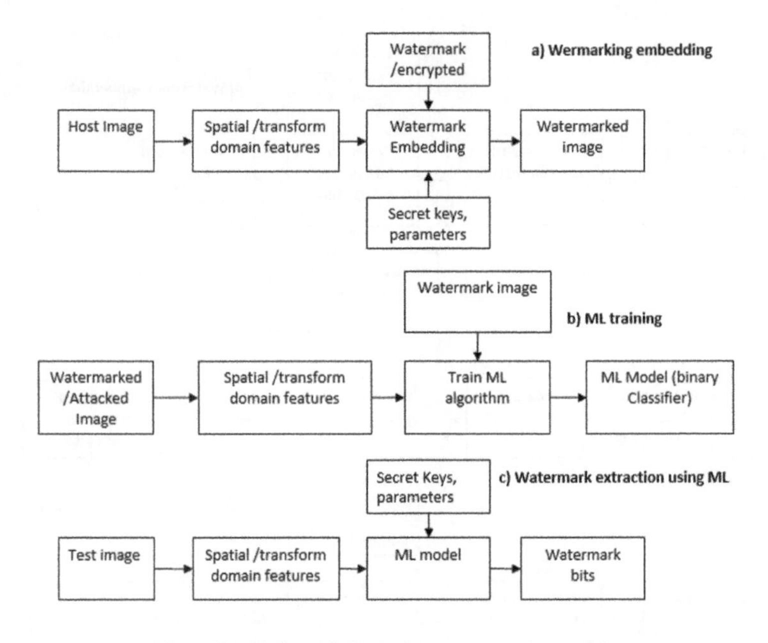

FIGURE 2.7 Type II ML based digital image watermarking.

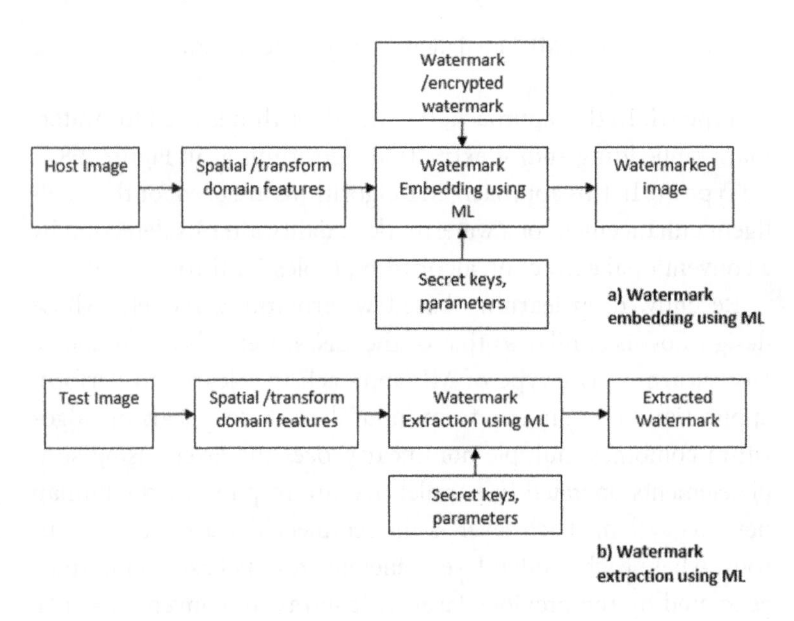

FIGURE 2.8 Type III ML based digital image watermarking.

TABLE 2.1 Comparison of Type I ML-Based Digital Image Watermarking Techniques

Sl. No.	Paper	Properties	Features	Embedding Method	Watermark Extraction	Advantages	Disadvantages
1	Zhang and Zhang (2004)	Spatial, blind Gray host (256 × 256)	Noise visibility function (NVF) is used in embedding process	Hopfield neural network is used for embedding watermark bits	Watermark bits are extracted using hash function and pseudorandom sequence	Simple to implement	Poor against watermarking attacks
2	Li et al. (2005)	Spatial, blind Gray host (512 × 512) Binary watermark (32 × 32)	Pseudorandom position changing	Addition of scaled watermark	The trained SVR estimates the watermark bits from the watermarked image	Robust to geometric attacks, such as blurring, rotation, scaling, and sharpening	Poor to JPEG, median filtering
3	Liu and Jiang (2006)	DCT Invisible Gray host (512 × 512) Binary watermark (64 × 64)	Radial basis function (RBF) neural network is used to finding optimal scaling factor	Addition of scaled watermark	Non-blind	Good robustness to cropping	Poor to LPF
4	Bansal and Bhadauria (2008)	Spatial, blind Gray host (512 × 512) Binary watermark (32 × 32)	Trained neural network weights are used as watermark bits	Modification of pixel values of host by addition of scaled watermark	Comparison of pixel values of watermarked image with trained neural network weights	Good robustness to geometric attacks	Poor to JPEG, filtering
5	Li et al. (2010)	DCT, non-blind Invisible Gray host (256 × 256) Binary watermark (32 × 32)	SVM is used for embedding watermark into host image	Addition of scaled watermark	Non-blind	Robust to attacks such as JPEG, noise addition, and Gaussian LPF	Less visual quality for watermarked image
6	Vafaei et al. (2013)	DWT, blind Invisible Gray host (512 × 512) Binary watermark (32 × 32)	3rd-level Haar DWT blocks	Feedforward neural network (FNN) is used to find optimal watermark strength	Using fuzzy logic	Robust to attacks such as JPEG, noise addition, median filter	Not reported

(Continued)

TABLE 2.1 (CONTINUED) Comparison of Type I ML-Based Digital Image Watermarking Techniques

Sl. No.	Paper	Properties	Features	Embedding Method	Watermark Extraction	Advantages	Disadvantages
7	Shareef and Fadel (2014)	Spatial, visible Color host (250 × 250)	Multilayered FNN is used for embedding watermark bits	Weight values (obtained by FNN) of host image are modified by watermark bits	Non-blind	Used to authenticate color image	Not reported
8	Rai and Singh (2018)	DWT + SVD Invisible Gray host (256 × 256) Binary Watermark (128 × 128)	SVM is used for separation of ROI and RONI of host image	Addition of scaled watermark	Non-blind	Good robustness to cropping	Poor to all other attacks
9	Zhou et al. (2018)	DCT Invisible Gray host (256 × 256) Binary watermark (32 × 32)	SVM is used for finding optimal embedding intensity and embedding locations	Addition of scaled watermark	Non-blind	JPEG compression, filtering, histogram equalization, cropping, and noise	Poor to rotations

TABLE 2.2 Comparison of Type II ML-Based Digital Image Watermarking Techniques

Sl. No.	Paper	Properties	Features	Embedding Method	Watermark Extraction	Advantages	Disadvantages
1	Zhang et al. (2002)	DWT, blind Invisible Gray host (512 × 512) Binary watermark (64 × 64)	L-level multiwavelet	By changing polarity and mean value of wavelet coefficients	Back-propagation neural network (BPNN)	JPEG, blurring, sharpening, and resampling	Not reported
2	Wang et al. (2006)	DWT, blind Invisible Gray host (512 × 512) Binary watermark (64 × 64)	L-level multiwavelet	Addition of scaled watermark	BPNN	JPEG, filtering, and cropping	Not reported
3	Tsai and Sun (2007)	Spatial, blind Invisible Color host (512 × 512) Binary watermark (64 × 64)	Pseudorandom position changing	Addition of scaled watermark	SVM	Geometric attacks such as blurring, rotation, scaling, and sharpening	Filtering, JPEG
4	Huang et al. (2008)	DWT, blind Invisible Gray host (512 × 512) Binary watermark (64 × 64)	DWT	Addition of scaled watermark	BPNN	Median filtering, low-pass filtering, cropping, scaling	Rotations, cropping, affine
5	Wu (2009)	DWT, non-blind Invisible Gray host (512 × 512) Watermark 64 bits	SVM	Addition of scaled watermark	SVM	JPEG, Gaussian noise	Filtering attacks, geometric attacks
6	Wen et al. (2009)	DWT, blind Invisible Gray host (512 × 512) Binary watermark (64 × 64)	2nd-level dual-tree wavelet transform (DTCWT)	Addition of scaled watermark	PNN	Additive white Gaussian noise, median filtering, pepper and salt noise, cropping, Gaussian noise	Affine attacks

(*Continued*)

TABLE 2.2 (CONTINUED) Comparison of Type II ML-Based Digital Image Watermarking Techniques

Sl. No.	Paper	Properties	Features	Embedding Method	Watermark Extraction	Advantages	Disadvantages
7	Peng et al. (2010)	DWT, blind Invisible Gray host (512 × 512) Binary watermark (32 × 32)	Single-level multiwavelet	Block mean value modulation	SVM	JPEG, low-pass filtering, noise addition, rotation, and scaling	Affine attacks
8	Jagadeesh et al. (2014)	DWT, blind Invisible Gray host (512 × 512) Binary watermark (64 × 32)	2nd-level DWT blocks	Modification of wavelet coefficients of blocks	SVM	Noise addition, filtering	Geometric attacks
9	Yahya et al. (2015)	DWT, blind Invisible Gray host (512 × 512) Binary watermark (64 × 64)	3rd-level Haar DWT blocks	Addition of scaled watermark	PNN	JPEG compression, rotation, Gaussian noise, cropping, and median filter	Not reported
10	Singh et al. (2016)	DWT, non-blind Invisible Color host (512 × 512) Grayscale watermark (128 × 128)	2nd- and 3rd-level DWT	Addition of scaled watermark	BPNN	Noise addition, filtering	Less visual quality for color watermarked image
11	Zear et al. (2017)	DWT, non-blind Invisible Color host (512 × 512) Color watermark (64 × 64)	3rd-level DWT	Addition of scaled watermark	BPNN	Noise addition, filtering	Less visual quality for color watermarked image

TABLE 2.3 Comparison of Type III ML-Based Digital Image Watermarking Techniques

Sl. No.	Paper	Properties	Features	Embedding Method	Watermark Extraction	Advantages	Disadvantages
1	Tsai et al. (2006)	DWT, blind Binary host (64 × 64)	ANN is used for generation of two random seeds	Wavelet coefficients	Comparison of coefficients of watermarked image with random seed value	Robust to all possible attacks	Visual information of scheme is not given
2	Yen and Wang (2006)	Spatial, blind Color host (512 × 512) Binary watermark (64 × 64)	SVM is used for finding information of embedding location in host image	Pixel values of host image are modified by watermark mask using neighboring average process	SVM	Good visual quality to color watermarked image	Not reported
3	Fu and Peng (2007)	DWT, non-blind Invisible	SVR is used to find best 2nd-level wavelet coefficients	Addition of scaled watermark	Extraction of watermark bits by comparison of modified wavelet coefficients with actual wavelet coefficients	Good visual quality to watermarked image	Poor to median filter
4	Shao et al. (2008)	DWT, blind Gray host (512 × 512) Binary watermark (32 × 32)	SVM is used for finding best 2nd-level wavelet coefficients of host image for embedding and of watermarked image for extraction	Addition of scaled watermark	Watermark bits extracted by comparing embedded pixel values and their nine-neighbor pixel values	Provide good robustness against all possible attacks	Less visual quality for watermarked image
5	Jain and Tiwari (2011)	DWT, blind Invisible Color host (256 × 256) Binary watermark (32 × 32)	3rd-level DWT blocks	Trained SVM is used for embedding watermark in wavelet coefficients of host blocks	Trained SVM is used for extraction of watermark from the wavelet coefficients of watermarked blocks	Provide good visual quality to watermarked image	Not reported

(Comtinued)

TABLE 2.3 (CONTINUED) Comparison of Type III ML-Based Digital Image Watermarking Techniques

Sl. No.	Paper	Properties	Features	Embedding Method	Watermark Extraction	Advantages	Disadvantages
6	Ramamurthy and Varadrrajan (2012)	DWT, blind Invisible Color host (256 × 256) Gray watermark (64 × 64)	4th-level DWT of blue component	Quantization and back-propagation neural network (BPNN)	Quantization and BPNN	Robust to salt and pepper noise, cropping, rotation	Cannot resist JPEG and median filter attacks
7	Ramamurthy and Varadrrajan (2012)	DWT, blind Invisible Color host (512 × 512) Gray watermark (64 × 64)	3rd-level DWT of blue component	Quantization and fuzzy logic	Quantization and fuzzy logic	Robust to cropping, JPEG compression, salt and pepper noise, and rotation attacks	Vulnerable to median filtering attack
8	Jagadeesh et al. (2013)	DWT, blind Invisible Gray host (512 × 512) Binary watermark (64 × 64)	2nd-level DWT	Trained SVM is used to find best wavelet coefficients for watermark embedding where watermark bits are inserted using scaling factor	Comparison of coefficients	Provides good visual quality to watermarked image	Cannot resist geometric attacks
9	Yahya et al. (2014)	Spatial, blind Gray host (256 × 256) Watermark 1024 bits	SVM is used to find best pixel locations in host image for embedding watermark bits	LSB substitution method	LSB extraction method	Provides good visual quality to watermarked image	Effect of scheme against attacks is not given

(Comtinued)

TABLE 2.3 (CONTINUED) Comparison of Type III ML-Based Digital Image Watermarking Techniques

Sl. No.	Paper	Properties	Features	Embedding Method	Watermark Extraction	Advantages	Disadvantages
10	Sharma and Kushwaha (2015)	DWT, blind Visible	Neural network is used for fusion of coefficients	Approximation wavelet coefficients of watermark bits are fused into approximation wavelet coefficients of host using neural network	Uses reverse process of embedding	Simple embedding process	Effect of scheme against attacks is not given
11	Jagadeesh et al. (2016)	DWT + DCT, blind Invisible Gray host (512 × 512) Gray watermark (32 × 32)	DCT	Fuzzy inference system + BPNN	Fuzzy inference system + BPNN	Robust to common attacks	Poor to median filtering
12	Mamatha and Venkatram (2016)	DWT, blind Invisible Gray host	Lifting wavelet transform	BPNN	BPNN	Provides good visual quality to watermarked image	Effect of scheme against attacks is not given
13	Kulkarni and Kuri (2017)	DWT + DCT, blind Invisible Gray host (512 × 512) Gray watermark (64 × 64)	3rd-level DWT + DCT	PNN	PNN	Robust to cropping, JPEG compression, salt and pepper noise, and rotation attacks	Vulnerable to affine

algorithms, where the extraction of input features and training of the model is done manually, deep learning does everything automatically. While ML models can be easily run over low-performance graphics processing units (GPUs), deep learning models demand high-performance GPUs for their operation. ML provides good results for small datasets, while deep learning techniques are good at dealing with very large datasets. The accuracy of the ML model is limited, while that of the deep learning model is broader.

The performance of classifiers designed using ML algorithms is measured by quality parameters such as accuracy, specificity, and sensitivity. The accuracy gives the probability that the classification test is correctly performed. The specificity gives the probability that the classification test has a negative value. The sensitivity gives the probability that the classification test has a positive value. The equations for these three parameters are given as

$$\text{Sensitivity}=\frac{\text{True\_Positive}}{\text{True\_Positive}+\text{False\_Negative}} \tag{2.7}$$

$$\text{Specificity}=\frac{\text{True\_Negative}}{\text{True\_Negative}+\text{False\_Positive}} \tag{2.8}$$

$$\text{Accuracy}=\frac{\text{True\_Positive}+\text{True\_Negative}}{\text{True\_Positive}+\text{True\_Negative}+\text{False\_Positive}+\text{False\_Negative}} \tag{2.9}$$

where False_Negative gives incorrectly classified negative values, False_Positive gives incorrectly classified positive values, True_Positive gives correctly classified positive values, and True_Negative gives correctly classified negative values.

The margins allowed for positives and negatives depend on the application. For example, in e-commerce applications, achieving zero false-positive rates is more important than achieving higher false-negatives rates. Plots of false-positive rate versus true-positive rate are referred to as receiver operating characteristic (ROC) curves, which are used to visualize the performance of a binary classifier.

2.12 CHALLENGES

Watermarking schemes embed watermarks for copyright protection of multimedia data, medical data, or biometric data. These schemes explore the various properties of signal processing transform to generate watermarked data. Every watermarking scheme has its own advantages and disadvantages, and its performance may vary with different types of host data. SVD-based watermarking schemes in combination with DWT are widely used and provide more transparency and robustness than all other watermarking schemes. Most of the watermarking in the literature is done by blind watermark extraction of the binary watermarks. The lower robustness, less security, less transparency, and less acceptability in practical applications are seen as weaknesses in many watermarking schemes. There are no benchmarking schemes available in the literature. Also, the higher computational time of some existing watermarking schemes is a limitation. Many simulation works are available for watermarking, but less work is done for hardware implementation of transform domain watermarking. Future research work can focus on these topics. The development of algorithms that deal with more general affine and geometric distortions needs to be addressed. Finding transformation-invariant domains for watermark embedding is required.

REFERENCES

Ansari, Rahim, Mrutyunjaya M. Devanalamath, K. Manikantan, and S. Ramachandran. "Robust digital image watermarking algorithm in DWT-DFT-SVD domain for color images." In *International Conference on Communication, Information & Computing Technology (ICCICT '12)*, pp. 1–6. IEEE, Piscataway, NJ, 2012.

Arena, Simone, Marcello Caramma, and Rosa Lancini. "Digital watermarking applied to MPEG-2 coded video sequences exploiting space and frequency masking." In *Proceedings of International Conference on Image Processing*, vol. 1, pp. 438–441. IEEE, Piscataway, NJ, 2000.

Ashour, Amira S., and Nilanjan Dey. "Security of multimedia contents: a brief." In *Intelligent Techniques in Signal Processing for Multimedia Security*, pp. 3–14. Springer, Cham, 2017.

Banerjee, Shubhendu, Sayan Chakraborty, Nilanjan Dey, Arijit Kumar Pal, and Ruben Ray. "High payload watermarking using residue number system." *International Journal of Image, Graphics and Signal Processing* 3: 1–8, 2015.

Bangaleea, R., and H. C. S. Rughooputh. "Performance improvement of spread spectrum spatial-domain watermarking scheme through diversity and attack characterisation." In *6th Africon Conference in Africa, 2002 (IEEE AFRICON)*, vol. 1, pp. 293–298. IEEE, Piscataway, NJ, 2002.

Bansal, Ashish, and Sarita Singh Bhadauria. "Watermarking using neural network and hiding the trained network within the cover image." *Journal of Theoretical & Applied Information Technology* 4(8): 663–670, 2008.

Baraniuk, Richard G. "Compressive sensing" [lecture notes]. *IEEE Signal Processing Magazine* 24(4): 118–121, 2007.

Bender, Walter, Daniel Gruhl, Norishige Morimoto, and Anthony Lu. "Techniques for data hiding." *IBM Systems Journal* 35(3–4): 313–336, 1996.

Bhatnagar, Gaurav, and Q. M. Jonathan Wu. "Biometrics inspired watermarking based on a fractional dual tree complex wavelet transform." *Future Generation Computer Systems* 29(1): 182–195, 2013.

Bhatnagar, Gaurav, Q. M. Jonathan Wu, and Balasubramanian Raman. "Biometric template security based on watermarking." *Procedia Computer Science* 2: 227–235, 2010.

Biswas, Debalina, Poulami Das, Prosenjit Maji, Nilanjan Dey, and Sheli Sinha Chaudhuri. "Visible watermarking within the region of non-interest of medical images based on fuzzy C-means and Harris corner detection." *Computer Science & Information Technology* 24: 161–168, 2013.

Borra, Surekha, and H. R. Lakshmi. "Visual cryptography based lossless watermarking for sensitive images." In *International Conference on Swarm, Evolutionary, and Memetic Computing*, pp. 29–39. Springer, Cham, 2015.

Candès, Emmanuel J. "Compressive sampling." In *Proceedings of the International Congress of Mathematicians*, vol. 3, pp. 1433–1452. Madrid, Spain, 2006.

Chakraborty, Sayan, Prasenjit Maji, Arijit Kumar Pal, Debalina Biswas, and Nilanjan Dey. "Reversible color image watermarking using trigonometric functions." In *International Conference on Electronic Systems, Signal Processing and Computing Technologies (ICESC)*, pp. 105–110. IEEE, Piscataway, NJ, 2014.

Chan, Chi-Kwong, and Lee-Ming Cheng. "Hiding data in images by simple LSB substitution." *Pattern Recognition* 37(3): 469–474, 2004.

Choudhary, Ankur, S. P. S. Chauhan, M. Afshar Alam, and Safdar Tanveer. "Schur decomposition and dither modulation: an efficient and robust audio watermarking technique." In *Proceedings of the CUBE International Information Technology Conference*, pp. 744–748. ACM, New York, 2012.

Datta, Sunanda, and Asoke Nath. "Data authentication using digital watermarking." *International Journal of Advance Research in Computer Science and Management Studies* 2(12): 30–45, 2014.

Deguillaume, Frederic, Gabriela Csurka, Joseph J. K. O'Ruanaidh, and Thierry Pun. "Robust 3D DFT video watermarking." In *Security and Watermarking of Multimedia Contents*, vol. 3657, pp. 113–125. International Society for Optics and Photonics, Bellingham, WA, 1999.

Dey, Nilanjan, Prasenjit Maji, Poulami Das, Shouvik Biswas, Achintya Das, and Sheli Sinha Chaudhuri. "An edge based blind watermarking technique of medical images without devalorizing diagnostic parameters." In *International Conference on Advances in Technology and Engineering (ICATE '13)*, pp. 1–5. IEEE, Piscataway, NJ, 2013.

Dey, Nilanjan, Shouvik Biswas, Poulami Das, Achintya Das, and Sheli Sinha Chaudhuri. "Lifting wavelet transformation based blind watermarking technique of photoplethysmographic signals in wireless telecardiology." In *World Congress on Information and Communication Technologies (WICT '12)*, pp. 230–235. IEEE, Piscataway, NJ, 2012.

Dili, Ruth Buse, and Elijah Mwangi. "An image watermarking method based on the singular value decomposition and the wavelet transform." In *AFRICON 2007*, pp. 1–5. IEEE, Piscataway, NJ, 2007.

Ding, Yuanyu, Xiaoshi Zheng, Yanling Zhao, and Guangqi Liu. "A video watermarking algorithm resistant to copy attack." In *Third International Symposium on Electronic Commerce and Security (ISECS '10)*, pp. 289–292. IEEE, Piscataway, NJ, 2010.

Donoho, David L. "Compressed sensing." *IEEE Transactions on Information Theory* 52(4): 1289–1306, 2006.

Durvey, Mohan, and Devshri Satyarthi. "A review paper on digital watermarking." *International Journal of Emerging Trends & Technology in Computer Science* 3(4): 99–105, 2014.

Ejima, Masataka, and Akio Miyazaki. "A wavelet-based watermarking for digital images and video." *IEICE Transactions on Fundamentals of Electronics, Communications and Computer Sciences* 83(3): 532–540, 2000.

Elbasi, Ersin. "Robust MPEG video watermarking in wavelet domain." *Trakya University Journal of Natural Sciences* 8(2): 87–93, 2007.

Essaouabi, A., and E. Ibnelhaj. "A 3D wavelet-based method for digital video watermarking." In *First International Conference on Networked Digital Technologies (NDT '09)*, pp. 429–434. IEEE, Piscataway, NJ, 2009.

Fakhr, M. "Robust watermarking using compressed sensing framework with application to MP3 audio." *International Journal of Multimedia & Its Applications (IJMA)* 4(6): 27–43, 2012.

Fan, Liang, and Fang Yanmei. "A DWT-based video watermarking algorithm applying DS-CDMA." In *TENCON, IEEE Region 10 Conference*, pp. 1–4. IEEE, Piscataway, NJ, 2006.

Fu, Gaoding, and Hong Peng. "Subsampling-based wavelet watermarking algorithm using support vector regression." In *EUROCON, 2007, the International Conference on "Computer as a Tool,"* pp. 138–141. IEEE, Piscataway, NJ, 2007.

Ganic, Emir, and Ahmet M. Eskicioglu. "Robust DWT-SVD domain image watermarking: embedding data in all frequencies." In *Proceedings of the 2004 Workshop on Multimedia and Security*, pp. 166–174. ACM, New York, 2004.

Gavini, Narasimha Swamy, and Surekha Borra. "Lossless watermarking technique for copyright protection of high resolution images." In *Region 10 Symposium, 2014 IEEE*, pp. 73–78. IEEE, Piscataway, NJ, 2014.

Gunjan, Reena, Priyanka Mitra, and Manoj S. Gaur. "Contourlet based image watermarking scheme using Schur factorization and SVD." In *International Conference on Advances in Communication, Network, and Computing*, pp. 337–340. Springer, Berlin, 2012.

Gupta, Akshya Kumar, and Mehul S. Raval. "A robust and secure watermarking scheme based on singular values replacement." *Sadhana* 37(4): 425–440, 2012.

Han, S. C., Z. N. Zhang, and Y. J. Zhang. "Robust zero-watermark algorithm based on non-subsampled shearlet transform and QR decomposition." *Journal of Optoelectronics Laser* 23(10): 1957–1964, 2012.

Hartung, Frank, and Martin Kutter. "Multimedia watermarking techniques." *Proceedings of the IEEE* 87(7): 1079–1107, 1999.

Hendriks, Richard C., Jesper Jensen, and Richard Heusdens. "Noise tracking using DFT domain subspace decompositions." *IEEE Transactions on Audio, Speech, and Language Processing* 16(3): 541–553, 2008.

Hernandez, Juan R., Martin Amado, and Fernando Perez-Gonzalez. "DCT-domain watermarking techniques for still images: detector performance analysis and a new structure." *IEEE Transactions on Image Processing* 9(1): 55–68, 2000.

Huang, Fangjun, and Zhi-Hong Guan. "A hybrid SVD-DCT watermarking method based on LPSNR." *Pattern Recognition Letters* 25(15): 1769–1775, 2004.

Huang, Song, Wei Zhang, Wei Feng, and Huaqian Yang. "Blind watermarking scheme based on neural network." In *7th World Congress on Intelligent Control and Automation (WCICA '08)*, pp. 5985–5989. IEEE, Piscataway, NJ, 2008.

Hussein, Jamal, and Aree Mohammed. "Robust video watermarking using multi-band wavelet transform." arXiv preprint arXiv: 0912.1826, 2009.

Jagadeesh, B., P. Rajesh Kumar, and P. Chenna Reddy. "Digital image watermark extraction in discrete wavelet transform domain using support vector machines." *International Journal on Recent Trends in Engineering & Technology* 11(1): 170, 2014.

Jagadeesh, B., P. Rajesh Kumar, and P. Chenna Reddy. "Robust digital image watermarking scheme in discrete wavelet transform domain using support vector machines." *International Journal of Computer Applications* 73(14): 1–7, 2013.

Jagadeesh, B., P. Rajesh Kumar, and P. Chenna Reddy. "Robust digital image watermarking based on fuzzy inference system and back propagation neural networks using DCT." *Soft Computing* 20(9): 3679–3686, 2016.

Jain, Yogendra Kumar, and Saurabh Tiwari. "An enhanced digital watermarking for color image using support vector machine." *International Journal of Computer Science and Information Technologies* 2(5): 2233–2236, 2011.

Kamlakar, M., Chhaya Gosavi, and A. Patankar. "Single channel watermarking for video using block based SVD." *International Journal of Advances in Computing and Information Researches* 1(2): 2012.

Kaushik, Awanish Kr. "A novel approach for digital watermarking of an image using DFT." *International Journal of Electronics and Computer Science Engineering* 1(1): 35–41, 2012.

Kulkarni, Gururaj, and Suresh Kuri. "Robust digital image watermarking using DWT, DCT and probabilistic neural network." In *International Conference on Electrical, Electronics, Communication, Computer, and Optimization Techniques (ICEECCOT '17)*, pp. 1–5. IEEE, Piscataway, NJ, 2017.

Lakshmi, H. R., and B. Surekha. "Asynchronous implementation of reversible image watermarking using mousetrap pipelining." In *IEEE 6th International Conference on Advanced Computing (IACC '16)*, pp. 529–533. IEEE, Bhimavaram, India, 2016.

Langelaar, Gerhard C., Iwan Setyawan, and Reginald L. Lagendijk. "Watermarking digital image and video data. A state-of-the-art overview." *IEEE Signal Processing Magazine* 17(5): 20–46, 2000.

Laur, Lauri, Pejman Rasti, Mary Agoyi, and Gholamreza Anbarjafari. "A robust color image watermarking scheme using entropy and QR decomposition." *Radioengineering* 24(4): 1025–1032, 2015.

Lee, Yeuan-Kuen, and Ling-Hwei Chen. "High capacity image steganographic model." *IEE Proceedings—Vision, Image and Signal Processing* 147(3): 288–294, 2000.

Li, Chun-Hua, Zheng-Ding Lu, and Ke Zhou. "An image watermarking technique based on support vector regression." In *IEEE International Symposium on Communications and Information Technology (ISCIT '05)*, vol. 1, pp. 183–186. IEEE, Piscataway, NJ, 2005.

Li, Jianzhong, Chuying Yu, B. B. Gupta, and Xuechang Ren. "Color image watermarking scheme based on quaternion Hadamard transform and Schur decomposition." *Multimedia Tools and Applications* 77(4): 4545–4561, 2017.

Li, Lei, Wen-Yan Ding, and Jin-Yan Li. "A novel robustness image watermarking scheme based on fuzzy support vector machine." In *3rd IEEE International Conference on Computer Science and Information Technology (ICCSIT '10)*, vol. 6, pp. 533–537. IEEE, Piscataway, NJ, 2010.

Lin, Ching-Yung, Min Wu, Jeffrey A. Bloom, Ingemar J. Cox, Matthew L. Miller, and Yui Man Lui. "Rotation, scale, and translation resilient watermarking for images." *IEEE Transactions on Image Processing* 10(5): 767–782, 2001.

Liu, Quan, and Xuemei Jiang. "Design and realization of a meaningful digital watermarking algorithm based on RBF neural network." In *International Conference on Neural Networks and Brain (ICNN&B '05)*, vol. 1, pp. 214–218. IEEE, Piscataway, NJ, 2005.

Lu, Chun-Shien, and Hong-Yuan Mark Liao. "Video object-based watermarking: a rotation and flipping resilient scheme." In *Proceedings of the 2001 International Conference on Image Processing*, vol. 2, pp. 483–486. IEEE, Piscataway, NJ, 2001.

Mamatha, P., and N. Venkatram. "Watermarking using lifting wavelet transform (LWT) and artificial neural networks (ANN)." *Indian Journal of Science and Technology* 9(17): 1–7, 2016.

Mansouri, A., A. Mahmoudi Aznaveh, F. Torkamani Azar. "SVD-based digital image watermarking using complex wavelet transform." *Sadhana* 34(3): 393–406, 2009.

Matsui, K. *Fundamentals of Digital Watermarks – New Protect Technology for Multimedia*. Morikita Publishing, Tokyo, Japan, 1998.

Meenakshi, K., Ch. Srinivasa Rao, and K. Satya Prasad. "A fast and robust hybrid watermarking scheme based on Schur and SVD transform." *International Journal of Research in Engineering and Technology* 3(4): 7–11, 2014.

Mehta, Rajesh, Navin Rajpal, and Virendra P. Vishwakarma. "LWT-QR decomposition based robust and efficient image watermarking scheme using Lagrangian SVR." *Multimedia Tools and Applications* 75(7): 4129–4150, 2016.

Mitra, Priyanka, Reena Gunjan, and Manoj S. Gaur. "A multi-resolution watermarking based on contourlet transform using SVD and QR decomposition." In *International Conference on Recent Advances in Computing and Software Systems (RACSS '12)*, pp. 135–140. IEEE, Piscataway, NJ, 2012.

Mohammad, Ahmad A. "A new digital image watermarking scheme based on Schur decomposition." *Multimedia Tools and Applications* 59(3): 851–883, 2012.

Mohan, B. Chandra, and K. Veera Swamy. "On the use of Schur decomposition for copyright protection of digital images." *International Journal of Computer and Electrical Engineering* 2(4): 781, 2010.

Mohan, B. Chandra, K. Veera Swamy, and S. Srinivas Kumar. "A comparative performance evaluation of SVD and Schur decompositions for image watermarking." In *IJCA Proceedings on International Conference on VLSI, Communications and Instrumentation (ICVCI)*, vol. 14, pp. 25–29. 2011.

Mostafa, Salwa A. K., A. S. Tolba, F. M. Abdelkader, and Hisham M. Elhindy. "Video watermarking scheme based on principal component analysis and wavelet transform." *International Journal of Computer Science and Network Security* 9(8): 45–52, 2009.

Naderahmadian, Yashar, and Saied Hosseini-Khayat. "Fast watermarking based on QR decomposition in wavelet domain." In *Sixth International Conference on Intelligent Information Hiding and Multimedia Signal Processing (IIH-MSP '10)*, pp. 127–130. IEEE, Piscataway, NJ, 2010.

Pal, Arijit Kumar, Nilanjan Dey, Sourav Samanta, Achintya Das, and Sheli Sinha Chaudhuri. "A hybrid reversible watermarking technique for color biomedical images." In *IEEE International Conference on Computational Intelligence and Computing Research (ICCIC '13)*, pp. 1–6. IEEE, Piscataway, NJ, 2013.

Pandhwal, Barun, and Devendra S. Chaudhari. "An overview of digital watermarking techniques." *International Journal of Soft Computing and Engineering (IJSCE)* 3(1): 416–420, 2013.

Peng, Hong, Jun Wang, and Weixing Wang. "Image watermarking method in multiwavelet domain based on support vector machines." *Journal of Systems and Software* 83(8): 1470–1477, 2010.

Potdar, Vidyasagar, Song Han, and Elizabeth Chang. "A survey of digital image watermarking techniques." In *3rd IEEE International Conference on Industrial Informatics (INDIN '05)*, pp. 709–716. IEEE, Piscataway, NJ, 2005.

Preda, Radu O., and Dragos N. Vizireanu. "Blind watermarking capacity analysis of MPEG2 coded video." In *8th International Conference on Telecommunications in Modern Satellite, Cable and Broadcasting Services (TELSIKS '07)*, pp. 465–468. IEEE, Piscataway, NJ, 2007.

Raghavendra, K., and K. R. Chetan. "A blind and robust watermarking scheme with scrambled watermark for video authentication." In *IEEE International Conference on Internet Multimedia Services Architecture and Applications (IMSAA '09)*, pp. 1–6. IEEE, Piscataway, NJ, 2009.

Rai, Ankur, and Harsh Vikram Singh. "Machine learning-based robust watermarking technique for medical image transmitted over LTE network." *Journal of Intelligent Systems* 27(1): 105–114, 2018.

Rajab, Lama, Tahani Al-Khatib, and Ali Al-Haj. "A blind DWT-SCHUR based digital video watermarking technique." *Journal of Software Engineering and Applications* 8(4): 224, 2015.

Rajab, Lama, Tahani Al-Khatib, and Ali Al-Haj. "Hybrid DWT-SVD video watermarking." In *International Conference on Innovations in Information Technology (IIT '08)*, pp. 588–592. IEEE, Piscataway, NJ, 2008.

Ramalingam, Mritha. "Stego machine–video steganography using modified LSB algorithm." *World Academy of Science, Engineering and Technology* 74: 502–505, 2011.

Ramamurthy, Nallagarla, and S. Varadarajan. "The robust digital image watermarking using quantization and fuzzy logic approach in DWT domain." arXiv preprint arXiv: 1302.4233, 2013.

Ramamurthy, Nallagarla, and S. Varadrrajan. "Robust digital image watermarking using quantization and back propagation neural network." *Contemporary Engineering Sciences* 5(3): 137–147, 2012.

Ramkumar, Mahalingam, Ali N. Akansu, A. Aydin Alatan. "A robust data hiding scheme for images using DFT." In *Proceedings of International Conference on Image Processing (ICIP '99)*, vol. 2, pp. 211–215. IEEE, Piscataway, NJ, 1999.

Rasti, Pejman, Salma Samiei, Mary Agoyi, Sergio Escalera, and Gholamreza Anbarjafari. "Robust non-blind color video watermarking using QR decomposition and entropy analysis." *Journal of Visual Communication and Image Representation* 38: 838–847, 2016.

Raval, M. S., and P. P. Rege. "Discrete wavelet transform based multiple watermarking scheme." In *TENCON 2003, Conference on Convergent Technologies for the Asia-Pacific Region*, vol. 3, pp. 935–938. IEEE, Piscataway, NJ, 2003.

Raval, M., M. Joshi, P. Rege, and S. Parulkar. "Image tampering detection using compressive sensing based watermarking scheme." In *Proceedings of MVIP 2011*. Pune, India, 2011.

Razafindradina, Henri Bruno, Nicolas Raft Razafindrakoto, and Paul Auguste Randriamitantsoa. "Improved watermarking scheme using discrete cosine transform and Schur decomposition." *International Journal of Computer Science and Network* 2(4): 25–31, 2013.

Santhi, V., and Arunkumar Thangavelu. "DWT-SVD combined full band robust watermarking technique for color images in YUV color space." *International Journal of Computer Theory and Engineering* 1(4): 424, 2009.

Saxena, Rahul, Nirupma Tiwari, and Manoj Kumar Ramaiya. "A survey work on digital watermarking." *IJLTEMAS* 4(5): 34–37, 2015.

Seddik, Hassen, Mounir Sayadi, and Farhat Fnaiech. "A new blind image watermarking method based on Shur transformation." In *Industrial Electronics 35th Annual Conference of IEEE (IECON '09)*, pp. 1967–1972. IEEE, Piscataway, NJ, 2009.

Serdean, C., M. Ambroze, M. Tomlinson, and G. Wade. "Combating geometrical attacks in a DWT based blind video watermarking system." In *Video/Image Processing and Multimedia Communications 4th EURASIP—IEEE Region 8 International Symposium*, pp. 263–266. Zadar, Croatia, 2002.

Shao, Yuanhai, Wei Chen, and Chan Liu. "Multiwavelet-based digital watermarking with support vector machine technique." In *Chinese Control and Decision Conference (CCDC '08)*, pp. 4557–4561. IEEE, Piscataway, NJ, 2008.

Shareef, Asmaa Qasim, and Roaa Essam Fadel. "An approach of an image watermarking scheme using neural network." *International Journal of Computer Applications* 92(1): 44–48, 2014.

Sharma, Anamika, and Ajay Kushwaha. "Image steganography scheme using neural network in wavelet transform domain." *International Journal of Scientific Engineering and Research* 3(10): 153–158, 2015.

Sheikh, Mona A., and Richard G. Baraniuk. "Blind error-free detection of transform-domain watermarks." In *IEEE International Conference on Image Processing (ICIP '07)*, vol. 5, p. V-453. IEEE, Piscataway, NJ, 2007.

Shoemaker, Chris. "Hidden bits: a survey of techniques for digital watermarking." *Independent Study, EER* 290: 1673–1687, 2002.

Singh, Amit Kumar, Basant Kumar, Sanjay Kumar Singh, S. P. Ghrera, and Anand Mohan. "Multiple watermarking technique for securing online social network contents using back propagation neural network." *Future Generation Computer Systems* 86: 926–939, 2016.

Singh, Priyanka, Balasubramanian Raman, and Partha Pratim Roy. "A multimodal biometric watermarking system for digital images in redundant discrete wavelet transform." *Multimedia Tools and Applications* 76(3): 3871–3897, 2017.

Solachidis, Vassilios, and Ioannis Pitas. "Optimal detector for multiplicative watermarks embedded in the DFT domain of non-white signals." *EURASIP Journal on Applied Signal Processing* 2004: 2522–2532, 2004.

Solachidis, Vassilios, and Loannis Pitas. "Circularly symmetric watermark embedding in 2-D DFT domain." *IEEE Transactions on Image Processing* 10(11): 1741–1753, 2001.

Soman, K. P. *Insight into Wavelets: From Theory to Practice*. PHI Learning Pvt. Ltd., New Delhi, 2010.

Song, Wei, Jian-Jun Hou, Zhao-Hong Li, and Liang Huang. "Chaotic system and QR factorization based robust digital image watermarking algorithm." *Journal of Central South University of Technology* 18(1): 116–124, 2011.

Sridevi, T., B. Krishnaveni, V. Vijaya Kumar, and Y. Rama Devi. "A video watermarking algorithm for MPEG videos." In *Proceedings of the 1st Amrita ACM-W Celebration on Women in Computing in India*, p. 35. ACM, New York, 2010.

Su, Qingtang, and Beijing Chen. "A novel blind color image watermarking using upper Hessenberg matrix." *AEU—International Journal of Electronics and Communications* 78: 64–71, 2017.

Su, Qingtang, Gang Wang, Xiaofeng Zhang, Gaohuan Lv, and Beijing Chen. "An improved color image watermarking algorithm based on QR decomposition." *Multimedia Tools and Applications* 76(1): 707–729, 2017.

Su, Qingtang, Yugang Niu, Gang Wang, Shaoli Jia, and Jun Yue. "Color image blind watermarking scheme based on QR decomposition." *Signal Processing* 94: 219–235, 2014.

Su, Qingtang, Yugang Niu, Hailin Zou, and Xianxi Liu. "A blind dual color images watermarking based on singular value decomposition." *Applied Mathematics and Computation* 219(16): 8455–8466, 2013.

Su, Qingtang, Yugang Niu, Xianxi Liu, and Yu Zhu. "Embedding color watermarks in color images based on Schur decomposition." *Optics Communications* 285(7): 1792–1802, 2012.

Su, Qingtang. "Novel blind colour image watermarking technique using Hessenberg decomposition." *IET Image Processing* 10(11): 817–829, 2016.

Surekha, B., and G. N. Swamy. "A spatial domain public image watermarking." *International Journal of Security and Its Applications* 5(1): 1–12, 2011.

Surekha, B., and G. N. Swamy. "Visual secret sharing based digital image watermarking." *International Journal of Computer Science Issues (IJCSI)* 9(3): 312, 2012a.

Surekha, B., and G. N. Swamy. "Digital image ownership verification based on spatial correlation of colors." In *Conference on Image Processing*, pp. 159–163. IET, London, UK, 2012b.

Thakkar, Falgun N., and Vinay Kumar Srivastava. "A blind medical image watermarking: DWT-SVD based robust and secure approach for telemedicine applications." *Multimedia Tools and Applications* 76(3): 3669–3697, 2017.

Thanki, Rohit M., and Ashish M. Kothari. "Digital watermarking: technical art of hiding a message." In *Intelligent Analysis of Multimedia Information*, pp. 431–466. IGI Global, Hershey, PA, 2017.

Thanki, Rohit M., Vedvyas J. Dwivedi, and Komal R. Borisagar. *Multibiometric Watermarking with Compressive Sensing Theory: Techniques and Applications*. Springer, Berlin, 2018.

Thanki, Rohit, Rushit Trivedi, Rahul Kher, and Divyang Vyas. "Digital watermarking using white Gaussian noise (WGN) in spatial domain." In *Proceeding of International Conference on Innovative Science and Engineering Technology (ICISET '11)*, vol. 1, pp. 38–42. V. V. P. Engineering College, Rajkot, India, 2011.

Thanki, Rohit, Surekha Borra, Vedvyas Dwivedi, and Komal Borisagar. "A steganographic approach for secure communication of medical images based on the DCT-SVD and the compressed sensing (CS) theory." *Imaging Science Journal* 65(8): 457–467, 2017.

Thanki, Rohit, Surekha Borra, Vedvyas Dwivedi, and Komal Borisagar. "A RONI based visible watermarking approach for medical image authentication." *Journal of Medical Systems* 41(9): 143, 2017.

Tiesheng, Fan, Lu Guiqiang, Dou Chunyi, and Wang Danhua. "A digital image watermarking method based on the theory of compressed sensing." *International Journal of Automation and Control Engineering* 2(2): 56–61, 2013.

Tsai, Hung-Hsu, and Duen-Wu Sun. "Color image watermark extraction based on support vector machines." *Information Sciences* 177(2): 550–569, 2007.

Tsai, Hung-Hsu, Chi-Chih Liu, and Kuo-Chun Wang. "Blind wavelet-based image watermarking based on HVS and neural networks." In *9th Joint International Conference on Information Sciences (JCIS '06)*. Atlantis Press, Paris, 2006.

Vafaei, M., H. Mahdavi-Nasab, and H. Pourghassem. "A new robust blind watermarking method based on neural networks in wavelet transform domain." *World Applied Sciences Journal* 22(11): 1572–1580, 2013.

Valenzise, Giuseppe, Marco Tagliasacchi, Stefano Tubaro, Giacomo Cancelli, and Mauro Barni. "A compressive-sensing based watermarking scheme for sparse image tampering identification." In *16th IEEE International Conference on Image Processing (ICIP '09)*, pp. 1265–1268. IEEE, Cairo, Egypt, 2009.

Van Loan, Charles F., and Gene H. Golub. JHU Press, Baltimore, 1996.

Wang, Junxiang, Ying Liu, and Yonghong Zhu. "Color image blind watermarking algorithm based on QR decomposition and voting in DWT domain." *International Journal of Security and Its Applications* 10(11): 33–46, 2016.

Wang, Zhenfei, Nengchao Wang, and Baochang Shi. "A novel blind watermarking scheme based on neural network in wavelet domain." In *The Sixth World Congress on Intelligent Control and Automation (WCICA '06)*, vol. 1, pp. 3024–3027. IEEE, Piscataway, NJ, 2006.

Wen, Xian-Bin, Hua Zhang, Xue-Quan Xu, and Jin-Juan Quan. "A new watermarking approach based on probabilistic neural network in wavelet domain." *Soft Computing* 13(4): 355–360, 2009.

Wu, Jianzhen. "A RST invariant watermarking scheme utilizing support vector machine and image moments for synchronization." In *Fifth International Conference on Information Assurance and Security (IAS '09)*, vol. 1, pp. 572–574. IEEE, Piscataway, NJ, 2009.

Yahya, Saadiah, Hanizan Shaker Hussain, and Fakariah Hani M. Ali. "DCT domain StegaSVM-Shifted LSB model for highly imperceptible and robust cover image." In *Proceedings of the 5th International Conference on Computing and Informatics,* ICOCI 2015, pp. 381–387. 11–13 August, Istanbul, Turkey, 2015.

Yahya, Al-Nabhani, Hamid A. Jalab, Ainuddin Wahid and Rafidah Md Noor. "Robust watermarking algorithm for digital images using discrete wavelet and probabilistic neural network." *Journal of King Saud University—Computer and Information Sciences* 27(4): 393–401, 2015.

Yen, Shwu-Huey, and Chia-Jen Wang. "SVM based watermarking technique." *Tamkang Institute of Technology* 9(2): 141–150, 2006.

Zear, Aditi, Amit Kumar Singh, and Pardeep Kumar. "Robust watermarking technique using back propagation neural network: a security protection mechanism for social applications." *International Journal of Information and Computer Security* 9(1–2): 20–35, 2017.

Zhang, Fan, and Hongbin Zhang. "Image watermarking capacity analysis using neural network." In *Proceedings of IEEE/WIC/ACM International Conference on Web Intelligence (WI'04)*, pp. 461–464. IEEE, Piscataway, NJ, 2004.

Zhang, Jun, Nengchao Wang, and Feng Xiong. "Hiding a logo watermark into the multiwavelet domain using neural networks." In *Proceedings of the 14th IEEE International Conference on Tools with Artificial Intelligence (ICTAI'02)*, pp. 477–482. IEEE, Piscataway, NJ, 2002.

Zhang, Xinpeng, Zhenxing Qian, Yanli Ren, and Guorui Feng. "Watermarking with flexible self-recovery quality based on compressive sensing and compositive reconstruction." *IEEE Transactions on Information Forensics and Security* 6(4): 1223–1232, 2011.

Zhou, Xiaoyi, Chunjie Cao, Jixin Ma, and Longjuan Wang. "Adaptive digital watermarking scheme based on support vector machines and optimized genetic algorithm." *Mathematical Problems in Engineering* 2018: 2685739, 2018.

CHAPTER 3

Watermarking Using Bio-Inspired Algorithms

MACHINE LEARNING (ML) ALGORITHMS, such as support vector regression (SVR), support vector machines (SVM), fuzzy logic, neural networks, and deep learning, are widely used in digital image watermarking for watermark embedding or extraction purposes. These techniques provide computational intelligence in identifying optimum intensities/locations of the host image to be modified and optimal scaling factors to be used during watermark embedding. Though the selection of ML algorithms partially depends on the type and size of images and the required output, there is no one machine learning algorithm fit to all requirements of applications related to sensitive image watermarking; therefore, finding the right ML algorithm is just a trial-and-error method. The performance of invisible, robust digital image watermarking depends on the imperceptibility and robustness, which in turn depends on factors such as watermark size and gain factors/scaling factors/features/locations chosen for watermark embedding.

In this chapter a variety of bio-inspired algorithms (Holland, 1975; Glover, 1977; Kirkpatrick et al., 1983; Koza, 1992, 1994; Kennedy and Eberhart, 1995; Dorigo et al., 1996; Storn and Price, 1997; Memon et al., 2018; Ghosh and Das, 2018) that are employed by digital image watermarking for optimum selection of parameters are discussed.

3.1 OPTIMIZATION AND ITS APPLICATION TO DIGITAL IMAGE WATERMARKING

Optimization is the science of identifying or choosing the best among a number of possible alternatives, after evaluating a number of possible solutions meeting one or more objective functions and associated assumptions (Powell, 1970). An optimization model can be defined as in Equation 3.1 (Venter, 2010):

Objective function: minimize or maximize $f(X)$

Subject to the **constraints**

$$a_j(Y) \geq 0, \quad j = 1, 2, \&, n;$$
$$b_j(Y) = 0, \quad j = n+1, n+2, \&, m \tag{3.1}$$

where:

Y is the vector of **decision variables**
$a(Y)$ are the **inequality constraints**
$b(Y)$ are the **equality constraints**

The optimization techniques are classified (Antoniou and Wu, 2007; Venter, 2010) based on:

- Type of constraints: Constrained and unconstrained optimization problems
- Nature of the equations involved: Linear, nonlinear, geometric, and quadratic
- Nature of design variables: Optimal control and non-optimal control problems

- Permissible value of the design variables: Integer and real-valued
- Number of objective functions: Single-objective and multi-objective
- Nature of problem: Deterministic and Stochastic Optimization problems

There are many approaches to find minima (local/global) other than gradient descent: (1) approximation methods (polynomial interpolation, Newton's method), (2) search methods (Dichotomous, Fibonacci, Golden-Section), (3) combination of (Davies, Swann, and Campey Algorithm), etc. The performance of the optimization algorithms is verified by collecting the mathematical convergence proofs and by testing over the benchmark problems. Random methods, such as random jumps, random walks, and simulated annealing, are quite inefficient, but can be used in initial stage to find the promising starting point and local minima and can be applicable to even non-differentiable functions; however, the results of random search methods are not repeatable. Other methods such as Nelder and Mead simplex method (Nelder and Mead, 1965), Powell's conjugate direction method (Powell, 1970) and cyclic coordinate search can be used as alternatives. The trend now is to use biologically inspired methods to get better accuracy in finding global minima, though they are complex. The algorithms covered in this chapter are:

- Genetic algorithm (GA)
- Genetic programming (GP)
- Ant colony optimization (ACO)
- Differential evolution (DE)
- Bee algorithm (BA)
- Bacterial foraging (BF)

- Cuckoo search algorithm (CSA)
- Cat swarm optimization (CSO)
- Particle swarm optimization (PSO)
- Firefly algorithm (FA)
- Tabu search (TS)
- Simulated annealing (SA)

3.2 IMAGE WATERMARKING USING GENETIC ALGORITHM (GA) AND GENETIC PROGRAMMING

The genetic algorithm (GA) (Baranowski and Eugene, 2018; Dey et al., 2015; Chatterjee et al., 2018), which evolved from Darwin's theory of natural evolution, is a direct search method and has the following advantages compared to traditional optimization methods:

- GA works on multiple point searches, while the traditional methods work on single point search.
- GA improves the global optimal value and provides more robustness.
- GA does not use any auxiliary value of problem parameters.
- GA is applicable to continuous or discrete optimization problems.
- GA uses probabilistic transfer function while traditional optimization methods use deterministic transfer function.

GA is suitable for unconstrained optimization problems but most of the problems are constrained in nature. As a first step, a constrained problem is converted into an unconstrained problem by adding an additional penalty function to the optimization problem based on the distance from the flexible region and number of constraints. The penalty function has to satisfy two requirements: (1) it is progressive, and (2) the factor of this function is the summarized value of all the loss done due to constraints violation.

GA uses a population of classes that gives optimal solutions. GA is naturally used for solving maximization problems. This algorithm is robust, flexible, and efficient on various types of problems. GA is not a simple random search optimization method, but it utilizes knowledge of previous iteration and generates new optimal solution. There are six steps in this algorithm: (1) problem identification, (2) initialization of class, (3) evaluation of fitness function, (4) constraint handling, (5) generation of new class, and (6) stopping criteria. The first step in applying GA is encoding, which sets its window limitation for use in the system. In GA, the information is represented by multiple chromosomes/genes, which are represented as a string of variables. The variables can be binary or real numbers and their length is determined by the problem specification. The two parameters, class and process, are initialized next. Each class is evaluated based on some fitness function measurement in the process of obtaining optimal solutions. The GA generates multiple class points with predefined size. This makes GA to search multiple different probabilities of the problem space and results in global optimal solution. After initialization of class, GA uses the survival principle of nature to search process and uses the fitness function as input information to determine the space for problem. The generation of new class is done using different operators such as selection, crossover, and mutation. The first step involves selection of a class from multiple classes according to the fitness function with respect to a given optimization problem. Once the selection process is over, the crossover, which is a recombination operator that combines sub information of two main chromosomes to produce new information, is applied. The mutation and crossover operators generate a large amount of data strings that may create two types of problems; (1) GA searches the entire space of the optimization problem due to less diversity in the initial data strings; (2) GA may have sub-optimum strings due to a wrong choice of initial class. These problems are overcome by the mutation operator in GA. This mutation operator is used to inject new genetic data into the genetic classes. In this process, the parent string can either replace the whole class or replace less fit value in string. During the operation of GA, the fitness function value increases gradually and at a particular condition, the increment in the fitness function value is not possible, which represents the optimal or near optimal solution. At this stage, the operation of GA is terminated. The basic flow chart of GA and a sample crossover and mutation operation are shown in Figures 3.1 and 3.2.

Genetic programming (GP) (Koza, 1994) is an extension of the genetic algorithm with differences in the output. While GA uses a direct encoding method to find the optimal solution, GP uses an indirect method. Another difference is that while GP uses a variable length encoding string, GA uses a fixed length encoding

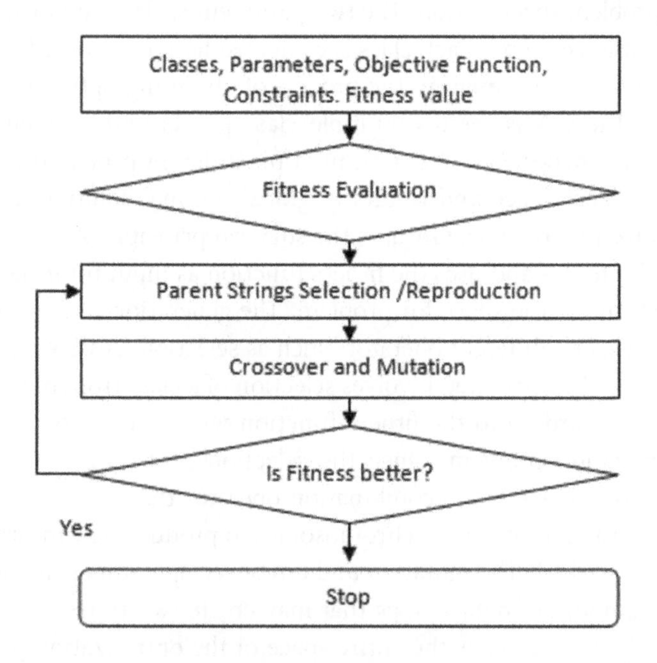

FIGURE 3.1 Basic flowchart of genetic algorithm (GA).

Before Crossover Operation	After Crossover Operation
Parent string 1: 0010 1010 Parent string 2: 1101 1101	Parent string 1: 0010 1101 Parent string 2: 1101 1010
Before Mutation Operation	**After Mutation Operation**
Parent string 1: 0010 1010 Parent string 2: 1101 1101	Parent string 1: 0110 1000 Parent string 2: 0101 1001

FIGURE 3.2 Crossover and mutation operation.

string. In a way, GP provides computer programming of GA. The steps involved in GP are as follows:

1. Generate an initial class of programs based on input function.
2. Execute each program in the class and assign a fitness value according to how to solve the optimization problem.
3. Create a new class of computer programs based on prior knowledge of existing programs using mutation and crossover operations.

In digital image watermarking, GA or GP is basically used to find optimization of scaling factors to be used in watermarking embedding and extraction process as shown in Figure 3.3. The watermark embedding into a host image and extraction from a test image are done iteratively using conventional watermarking algorithms and GP/GA generated scaling factors/locations/intensities. In each iteration, the corresponding imperceptibility and robustness are calculated and substituted in the fitness function to meet the stopping criteria of the watermarking optimization problem. The stopping criterion is when the fitness value remains stable with iterations, or if the number of iterations meets its limit. The final value of the fitness function once stopping criteria are met is considered the optimized scaling factor for that particular host image. The test images may even be a set of watermarked images affected by various attacks.

A variety of fitness functions, which are a function of various performance metrics, are to be used with the GA-based image

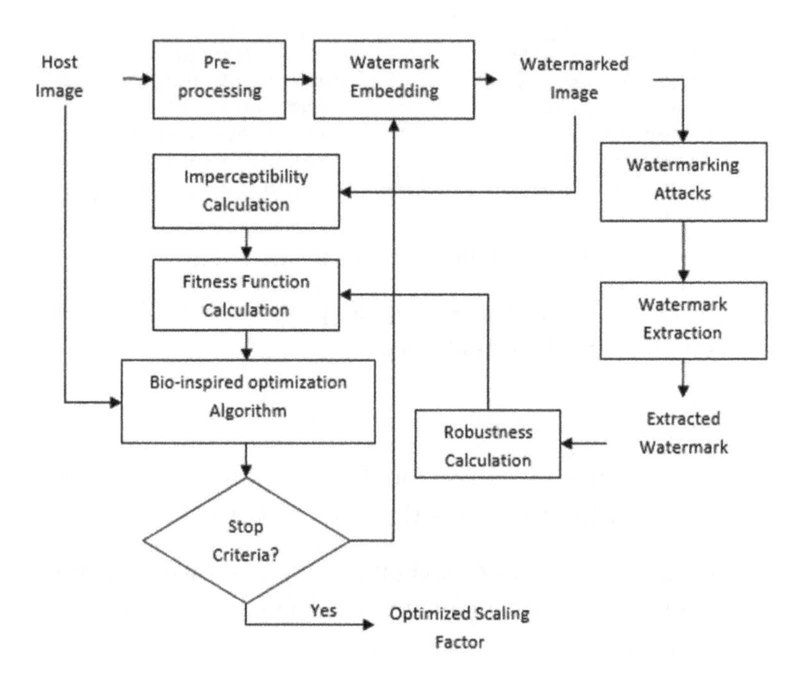

FIGURE 3.3 Image watermarking using evolutionary methods.

watermarking. Table 3.1 compares various GA-/GP-based digital image watermarking techniques with respect to their fitness functions, embedding domains and types of watermark detection. It is to be noted that the GA-based watermarking approaches are combined with various image processing transforms, such as discrete wavelet transform (DWT), singular value decomposition (SVD), discrete cosine transform (DCT), vector quantization (VQ), etc., where the performance of each watermarking technique depends on the nature of host image, embedding process, extraction process, and the number of iterations performed in GA or GP. The imperceptibility parameter (PSNR) in existing GA-based watermarking techniques varies between 30 dB and 60 dB, while the average value of the robustness parameter (NC) is between 0.4 and 1. Most of the existing techniques are defined for grayscale host images. The computational time of these techniques is very

TABLE 3.1 Comparison of GA/GP-Based Digital Image Watermarking Algorithms

Reference Paper	GA Used for	Domain	Type of Detection	Fitness Function
Shieh et al. (2004)	Finding optimal scaling factor	DCT	Blind and comparison of coefficients	$f = \text{PSNR} + \lambda \cdot \sum_{i=1}^{N} \text{NC}\left(w, w_i^*\right)$
Huang et al. (2007)	Finding optimal scaling factor	DCT, DWT	Blind and comparison of coefficients	$f = \text{PSNR} + \lambda_1 \cdot \frac{1}{N} \sum_{i=1}^{N} \text{BCR}\left(w, w_i^*\right) + \lambda_2 \cdot \text{Bits}_N$
Chu et al. (2008)	Finding optimal scaling factor	DWT	Blind and comparison of coefficients	$f = \text{PSNR} + \lambda \cdot \text{BCR}(w, w^*)$
Aslantas (2008)	Finding optimal scaling factor	SVD	Non-blind	$f = \left[1 \Big/ \left(\frac{1}{N} \sum_{i=1}^{N} \text{corr}\left(w, w_i^*\right) \right) - \text{corr}\left(I, \text{WI}\right) \right]$
Maity et al. (2013)	Finding best wavelet coefficients for embedding of watermark	DWT	Blind and comparison of coefficients	$f = \frac{3\text{MSSIM} \times (1-\text{PG}) \times \text{CC}}{\left[\text{MSSIM} + (1-\text{PG})\text{CC} + \text{MSSIM}(1-\text{PG})\right]}$
Nurdin et al. (2017)	Finding best wavelet coefficients for embedding of watermark	DWT	Blind and comparison of coefficients	$f = \frac{\sum_{c=1}^{3} \text{PSNR}_c}{3} - \sum_{i=1}^{22} \left(1 - \text{NC}_i\right) + 5$

Where, f, fitness function; NC, normalized cross-correlation; corr, correlation; BCR, bit correct rate; MSSIM, multiscale structural similarity index measure; CC, correlation coefficients; PG, process gain; λ, constant; N, number of attacks; w, original watermark image; w^*, recovered watermark image; I, original host image; WI, watermarked image.

high (around 2 minutes), which is the major limitation of image watermarking techniques using GA or GP.

3.3 IMAGE WATERMARKING USING DIFFERENTIAL EVOLUTION (DE)

Differential evolution (DE) is based on stochastic population (Storn and Price, 1997) and follows the basic flow of genetic algorithm with some difference in mutation and selection operations. Only two parameters, mutation factor and probability of crossover, are required to tune the algorithm. DE selects random vectors and then takes difference of the vectors. The mean of the vectors' difference is taken as parent vector with a special operator in finding searching space. The main difference between DE and GA is that mutation operation automatically takes increments to the best optimal value based on the evolutionary process stage. A few features of DE are as follows:

1. DE is rarely used for solution of local optimization as it searches for the global optimal solution by manipulating a class of objective functions and searches multiple areas within the search space.
2. DE is mainly used for solving optimization problems with non-smooth input functions.
3. DE allows for floating-point input data, manipulated data, and output data.
4. DE uses arithmetic operation for searching of space.
5. DE is not usable for large sized data.

In DE, the optimal solutions are represented in terms of floating-point numbers. The mutation operation is also different from GA. The weighted difference between two parent strings is added with a third string to generate the optimal solution, followed by a crossover operation to combine the optimal solution of the mutation operation with the optimal target solution. Then, a selection operation is applied to compare the fitness function value of operations solutions, the target solution, and the trail solution to decide which solution is to carry forward in the next stage. The basic DE algorithm has four steps: (1) initialization of class, (2) mutation operation, (3) crossover operation, and (4) selection operation. The main advantages of DE are that it is easy to implement, requires fewer parameters, and has faster convergence when compared to GA or GP. The main limitation of this algorithm is that its performance may be affected by noise due to its greedy nature. Table 3.2 compares various DE-based digital image watermarking techniques with respect to their fitness functions, embedding domain, and type of watermark detection.

The watermarking approaches based on DE are mainly implemented using singular value decomposition (SVD) in transform domain (Ali et al., 2014a; Aslantas, 2009). The performance of each available watermarking technique depends on the nature of the host image, the SVD-based embedding process, the extraction process, and the number of iterations performed in DE. The imperceptibility parameter (PSNR) values achieved with such existing watermarking techniques varies between 32 to 50 dB, while normalized correlation (NC) varies between 0.95 and 1. The existing techniques are defined only for grayscale images. The computational time of the existing techniques is not mentioned in the literature. The lack of recorded data is one of the limitations of watermarking using DE.

3.4 IMAGE WATERMARKING USING SWARM ALGORITHMS

Swarm algorithms are the most recent and widely used bio-inspired algorithms for finding optimized solutions. They are extensions of evolutionary computational techniques. While GA, GP, and DE are based on genetic adaptations of class, swarm algorithms are based on social behavior of class. The word "swarm" comes from irregular movement of an object in the problem space. These

TABLE 3.2 Comparison of DE-Based Digital Image Watermarking Algorithms

Reference Paper	Domain + Techniques Used	Type of Detection	Similarity Measures Used	Fitness Function
Peng et al. (2014)	Ridgelet Transform	Blind and Comparison of Coefficients	BER	$f = \text{PSNR} + \lambda \cdot \sum_{r=1}^{R} (1 - \text{BER}_r)$
Ali et al. (2014a)	DCT + SVD	Blind and Comparison of Coefficients	NC	$f = \frac{N}{\sum_{i=1}^{N} \text{NC}(w, w_i^*)} - \text{NC}(I, \text{WI})$
Aslantas (2009)	SVD	Non-blind	Corr	$F_i = \left[1 \Big/ \left(\frac{1}{t} \sum_{i=1}^{t} \text{corr}_w (w, w_i^*) - \text{corr}_w (I, \text{WI}) \right) \right]$ $f_i = \frac{1}{F_i}$

Where, f, fitness function; NC, normalized cross-correlation; corr, correlation; BER, bit error rate; w, original watermark image; w^*, recovered watermark image; I, original host image; WI, watermarked image.

algorithms are called trajectory tracking algorithms and are described by five fundamental principles (Waleed et al., 2014):

- Proximity: the class should have simple space and computational time
- Quality: the class should respond to the quality factors within the operating space
- Diverse Response: the class should perform equally in different operating spaces
- Stability: the behavior of class should remain unchanged if operating space is changing
- Adaptability: the class should change its operating behavior according to the computational time of algorithm

Various types of swarm algorithms are available in the literature (Venter, 2010; Binitha and Siva Sathya, 2012). These algorithms are

- ant colony optimization (ACO)
- bacterial foraging (BF)
- bee algorithm (BA)
- cat swarm optimization (CSO)
- cuckoo search algorithm (CSA)
- firefly algorithm (FA)
- particle swarm optimization (PSO)

In image watermarking, all these algorithms are used for optimization of scaling factor.

3.4.1 Image Watermarking Using Ant Colony and Bee Colony

Ant colony optimization (ACO) is the most widely used and best swarm algorithm (Dorigo et al., 1996) and is based on the movements of ants in the jungle. It refers to indirect communication approach within self-organization system via each class modifying their own environment. By modeling and simulating various behavior of ants, such as foraging, brood sorting, nest building, etc., these algorithms can solve complex optimization problems. ACO has three main functions to perform: (1) construction of solution process by moving artificial ants through adjacent states of problem space using standard transition rules: (2) updating of pheromones which update one complete optimal solution: (3) some additional actions with regard to global perspective. The advantages of ACO are that it provides inherent parallelism and rapid solution and that it has many dynamic applications. The limitations of this algorithm are that it is difficult for theoretical analysis of large scale of ant populations and probability distribution of function changes by iteration. Very few watermarking techniques based on ACO algorithm and transforms, such as fast Fourier transform (FFT) and discrete wavelet transform (DWT), are developed (Loukhaoukha et al., 2011; Al-Qaheri et al., 2010). The average PSNR achieved with existing ACO-based watermarking techniques is around 35 to 54 dB. The computational time of these existing techniques is not mentioned in the literature. The payload capacity of these existing techniques is less.

Bee algorithm (BA) is inspired by different characteristics of honey bees, such as movement or dancing (foraging behavior) of honey bees in nature. In virtual bee algorithm (Yang, 2005) pheromone concentrations are linked directly with optimization problem function. The artificial bee colony (ABC) algorithm (Karaboga, 2005) considers three types of honey bees: onlooker bees (observer bees), employed bees (forager bees), and scouts, all of which are available in the colony. This algorithm uses the transmission ability of a bee in order to find optimum locations or paths. The optimization problem is related to finding optimal routes and paths and is solved by a combination of ant and bee algorithms. This algorithm is very flexible and effectively deals with discrete optimization problems.

In digital image watermarking, ACO and ABC algorithms are mainly used to find optimized scaling factors based on the intensity of bee interactions. The solution will be updated after each iteration of ABC. As a first step initial solutions/population (scaling factors) are randomly generated. Each member vector in a population denotes a possible solution to the problem. The fitness function associated with PSNR and NC is evaluated for imperceptivity and robustness. There are very few works reported on image watermarking techniques based on bee algorithm (BA) for optimization (Farhan et al., 2011; Chen et al., 2012; Lee et al., 2014). These techniques are implemented using DCT, DWT, and SVD. The average PSNR is around 29 dB and the average payload capacity is around 0.0625 bpp, which is much less. Table 3.3 compares various ACO/ABC/BA-based digital image watermarking techniques with respect to their fitness functions, embedding domains, and type of watermark detection.

3.4.2 Image Watermarking Using Cuckoo Search Algorithm

Cuckoo search algorithm (CSA) (Yang and Deb, 2009) is a class-based algorithm like GA and is inspired by brood parasitic behavior of cuckoo birds and Lévy flights. There are three ideal rules that describe basic cuckoo algorithm; (1) each cuckoo gives one egg at a time and put its egg in randomly chosen nest; (2) the best nests with high numbers of eggs will carry over to the next generation; (3) the number of host nests is fixed and the probability that the host bird discovers egg lies in the interval [0, 1]. In this case, the host bird can either dispose of the nest or throw the egg.

In digital image watermarking, CSA is mainly used to find optimized scaling factors. The steps involved in image watermarking using CSA are given as follows (Dey et al., 2013):

1. Randomly select a set of N solutions called nests, each having individual gain factor values within a specific range.

TABLE 3.3 Comparison of ACO/ABC/BA-Based Digital Image Watermarking Algorithms

Reference Paper	Domain + Techniques Used	BIA and Its Used	Type of Detection	Similarity Measures Used	Fitness Function
Loukhaoukha et al. (2011)	LWT + SVD	Ant Colony Optimization (ACO) and used to find optimal scaling factor	Non-blind	NC	$F(X)=\begin{pmatrix} \frac{1}{\mathrm{NC}(I,\mathrm{WI})} \\ \frac{1}{\mathrm{NC}(w,w^{*})} \\ \frac{1}{\mathrm{NC}(w,w_1^{*})} \\ \vdots \\ \frac{1}{\mathrm{NC}(w,w_N^{*})} \end{pmatrix}$ $F_{\mathrm{obj}}(X)=\sum_{i=1}^{T+2}\left(e^{p\cdot\omega}-1\right)\cdot e^{P(F(X)-F_0)}$

(*Continued*)

TABLE 3.3 (CONTINUED) Comparison of ACO/ABC/BA-Based Digital Image Watermarking Algorithms

Reference Paper	Domain + Techniques Used	BIA and Its Used	Type of Detection	Similarity Measures Used	Fitness Function
Al-Qaheri et al. (2010)	FFT	ACO and used to find best FFT coefficients of host image	Blind	Not reported	Not reported
Chen et al. (2012)	DWT+SVD	Artificial Bee Colony (ABC) and used to find optimal scaling factor	Non-blind	Corr	$F = \min\left\{ N \Big/ \left(\sum_{i=1}^{N} \text{corr}_{w}\left(w, w_i^*\right) - \text{corr}_I(I, \text{WI}) \right) \right\}$
Lee et al. (2014)	DCT	ABC and used to find best DCT coefficients of host image	Blind	NC	$F = \bigcup_{m=1}^{M/8} \bigcup_{n=1}^{N/8} \left\{ F_{(m,n)}(i)\Big\vert_{i=1}^{\text{NUM}_w} = \text{RS}(D_{(m,n)}(k)\Big\vert_{k=1}^{63} \right\}$

Where, F, fitness function; NC, normalized cross-correlation; w, original watermark image; w^*, recovered watermark image; I, original host image; WI, watermarked image; RS (), random selection operator that randomly selects NUMw frequency bands to embed w from the 63 AC coefficients in each DCT blocks, p, ω and F_0 = positive constants

2. Perform watermarking using each solution set and calculate corresponding fitness value.
3. Store the solution set with the best fitness value as the best solution for scaling factors.
4. Get new cuckoos/solutions by Lévy flight and compare with the existing solutions. If the present cuckoo is better than the previous one, keep it and update the best fitness value.
5. Abandon a fraction (*Pa*) of inferior solutions and generate new solutions. Keep the best solutions update the best fitness value.
6. Repeat steps 3 to 5 i number of times.

Table 3.4 compares various CSA-based digital image watermarking techniques with respect to their fitness functions, embedding domains, and type of watermark detection.

3.4.3 Image Watermarking Using Particle Swarm Optimization

Particle swarm optimization (PSO) algorithm is based on the social behavior of the schooling of fish/flocking of birds. Initially, a class of particles is generated randomly and then the optimal value is calculated using an iterative search method. A velocity vector as well as position vector is calculated for each particle at every iteration followed by local best fitness value and global best calculations. The best particle from all the local best particles gives the global optimal particle. The basic steps of PSO algorithm are listed as follows (Chakraborty et al., 2013):

1. Generate N number of initial particle positions randomly within a specific range.
2. Find the local best solution of each particle by applying the initial value of the particle to watermark the image and extract the original image.

TABLE 3.4 Comparison of CS-Based Digital Image Watermarking Algorithms

Reference Paper	Domain + Techniques Used	Type of Detection	Similarity Measures Used	Fitness Function
Dey et al. (2013)	DWT	Blind and using correlation of PN sequences	NC	$f = \text{PSNR} + 100 \times \text{NC}$
Ali et al. (2014b)	RDWT + SVD	Non-blind	NC	$f = \frac{N}{\sum_{i=1}^{N} \text{NC}\left(w, w_i^*\right)} - \text{NC}(I, \text{WI})$
Ali and Ahn (2018)	DWT	Non-blind	NC	$f = 10 \times \left\lvert \text{PSNR} - \text{PSNR}_{\text{target}} \right\rvert + \left(1 - \frac{1}{N} \sum_{i=1}^{N} \text{NC}_i\right)$
Waleed et al. (2015)	DWT	Blind and comparison of coefficients	Not reported	$f = \max \text{PSNR}$
Mishra and Agarwal (2016)	DWT + SVD	Non-blind	NC	$f = \text{PSNR} + \lambda \left[\text{NC}\left(w, w^*\right) + \sum_{i=1}^{t} \text{NC}\left(w, w_i^*\right) \right]$
Issa (2018)	RDWT + SVD	Non-blind	NC	$f = \frac{N}{\sum_{i=1}^{N} \text{NC}\left(w, w_i^*\right)} - \text{NC}(I, \text{WI})$

Where, f, fitness function; NC, normalized cross-correlation; w, original watermark image; w^*, recovered watermark image; I, original host image; WI, watermarked image.

3. Change the particle velocity.
4. Find the best solution and update the global best position if the present position is better than the previous global best.
5. Change the particle position according to global best.
6. Repeat from step 3 until maximum iterations are achieved.

The fitness functions are calculated using different evaluation parameters of watermarking which are related to quality and robustness. According to the literature, the PSO algorithm is not only used for calculation of optimal scaling factor but also for selection of the best frequency coefficients or pixel locations of a host image for watermark embedding (Li and Wang, 2007; Bedi et al., 2012). Tables 3.5 and 3.6 compare various PSO-based digital image watermarking techniques with respect to their fitness functions, embedding domains, and type of watermark detection. The PSO is used along with various image processing transforms, such as DCT, DWT, and SVD, and is applied on different types of images, such as grayscale, color, and medical. The limitations of these techniques are that they have less imperceptibility and less payload capacity.

3.4.4 Image Watermarking Using Firefly Algorithm

Firefly algorithm (FA) (Yang et al. 2012) is inspired by behavior of fireflies which produce short and rhythmic flashing light. The algorithm is based on three constraints; (1) all fireflies are unisex and will be attracted to each other based on their sex; (2) the attractiveness of fireflies depends on their brightness of light (the firefly with the low light moves toward the firefly with the brighter light); (3) the brightness of the firefly is determined by the optimization problem. The firefly attractiveness Y is given by:

$$Y = Y_0 e^{-\gamma n^2} \tag{3.2}$$

TABLE 3.5 Comparison of PSO-Based Digital Image Watermarking Algorithms (Finding Best Embedding Locations)

Reference Paper	Domain + Techniques Used	Type of Detection	Similarity Measures Used	Fitness Function
Li and Wang (2007)	DCT	Semiblind	Not reported	$f = \mathrm{PSNR}(I, \mathrm{WI})$
Aslantas et al. (2008)	DCT	Blind and comparison of coefficients	NC	$f = \left[1 \Big/ \left(\frac{1}{N} \sum_{i=1}^{N} \mathrm{corr}\left(w, w_i^*\right) \right) - \mathrm{corr}(I, \mathrm{WI}) \right]^{-1}$
Rohani and Avanaki (2009)	DCT	Non-blind	SSIM	$f = 1 - \mathrm{SSIM}(I, \mathrm{WI})$
Findik et al. (2010)	Spatial	Blind	BCR	$f = \sqrt{\sum_{i=1}^{N} \left(\mathrm{Test}_i - \mathrm{Train}_i\right)^2}$
Fakhari et al. (2011)	DWT	Non-blind	SSIM and NC	$f = \left(\frac{100}{\mathrm{PSNR}} \right) + 7 \times (1 - \mathrm{SSIM}) + 2 \times \sum_{Q=40\%}^{100\%} \left(N_{w,\mathrm{jpg}} - N_{w^*,\mathrm{jpg}} \right) + 1.5 \times \sum_{\alpha=1}^{4} \left(N_{w,Rsz} - N_{w^*,Rsz} \right)$

(*Continued*)

TABLE 3.5 (CONTINUED) Comparison of PSO-Based Digital Image Watermarking Algorithms (Finding Best Embedding Locations)

Reference Paper	Domain + Techniques Used	Type of Detection	Similarity Measures Used	Fitness Function
				$+2\times\sum_{\alpha=1}^{2}\left(N_{w,\text{Noise}}-N_{w^*,\text{Noise}}\right)$ $+1.5\times\sum_{\alpha=1}^{4}\left(N_{w,\text{Rot}}-N_{w^*,\text{Rot}}\right)$ $+\left(N_{w,crp}-N_{w^*,crp}\right)$
Wang et al. (2011)	DWT	Non-blind	NC	$f=\text{PSNR}/100+\sum_{i=1}^{3}\text{NC}_i$
Wu et al. (2011)	DCT	Blind and comparison of coefficients	NC	$f=\text{PSNR}+\sum_{i=1}^{3}\lambda_i\cdot\text{NC}_i$
Bedi et al. (2012)	DCT	Blind and comparison of coefficients	SSIM and BER	$f=\lambda(1-\text{SSIM}(I,\text{WI}))+(1-\lambda)\text{BER}\left(w,w^*\right)$
Bedi et al. (2013)	Spatial	Blind and comparison of coefficients	SSIM and BER	$f=\lambda\left(1-\text{SSIM}(I,\text{WI})\right)+(1-\lambda)\text{BER}\left(w,w^*\right)$

Where, *f*, fitness function; SSIM, structural similarity index measure; NC, normalized cross-correlation; *corr*, correlation; BER, bit error rate; λ, constant value; *N*, number of attacks; *w*, original watermark image; *w**, recovered watermark image; *I*, original host image; WI, watermarked image.

TABLE 3.6 Comparison of PSO-Based Digital Image Watermarking Algorithms (Finding Optimal Scaling Factors)

Reference Paper	Domain	Type of Detection	Similarity Measures Used	Fitness Function
Vellasques et al. (2011)	Spatial	Blind	BCR	$f = we_1 \cdot \mathrm{BCR}^{-1} + we_2 \cdot \mathrm{DRDM}$
Run et al. (2012)	DCT + SVD	Non-blind	NC	$f = \frac{\max\left[\mathrm{corr}\left(w,w^*\right)+\mathrm{corr}\left(I,\mathrm{WI}\right)\right]}{2}$
Vellasques et al. (2013)	Spatial	Blind	BCR	$f = \max_{i=1,\ldots,i}\left\{\left(1-we_i\right)\left(\lambda \cdot \mathrm{DRDM}-r_i\right)\right\}$
Golshan and Mohammadi (2013)	DCT + DWT + SVD	Non-blind	NC	$f = \mathrm{PSNR} + \lambda \sum_{i=1}^{r} \mathrm{NC}_i$
Peng et al. (2014)	Ridgelet transform	Blind	BER	$f = \mathrm{PSNR} + \lambda \cdot \sum_{i=1}^{N}\left(1-\mathrm{BER}_i\right)$
Li (2014)	Gyrator transform	Non-blind	NC	$f = \mathrm{PSNR}/100 + 1.1 \times \left(\sum_{i=1}^{5} \mathrm{NC}_i\right)/5$
Verma et al. (2016)	DWT + SVD	Non-blind	NC	$f = \frac{\max\left[\mathrm{corr}\left(w,w^*\right)+\mathrm{corr}(I,\mathrm{WI})\right]}{2}$

(Continued)

TABLE 3.6 (CONTINUED) Comparison of PSO-Based Digital Image Watermarking Algorithms (Finding Optimal Scaling Factors)

Reference Paper	Domain	Type of Detection	Similarity Measures Used	Fitness Function
Rao et al. (2017)	IWT + SVD	Blind and comparison of coefficients	SSIM and NC	$f = 1 - \text{Average(NC)}$
Thakkar and Srivastava (2017)	DWT + SVD	Blind and comparison of coefficients	NC	$f = \frac{\max\left[\text{corr}\left(w, w^{*}\right) + \text{corr}(I, \text{WI})\right]}{2}$
Sanku et al. (2018)	DWT + SVD	Non-blind	NC	$f = Q + \text{PSNR} + \frac{1}{N}\sum_{i=1}^{N} \lambda_i \left(100 \times \text{NC}_i\right)$

Where, f, fitness function; NC, normalized cross-correlation; corr, correlation; DRDM, distance reciprocal distortion measure; BER, bit error rate; Q, image quality index factor; λ, constant value; N, number of attacks; w, original watermark image; w^*, recovered watermark image; I, original host image; WI, watermarked image.

Where, Y_0 is the attractiveness value at $n = 0$.

The movement of a firefly a toward another brighter firefly b is calculated as:

$$x_a^{t+1} = x_a^t +_0 e^{-r_{ab}^2} \left(x_b^t - x_a^t \right) + \alpha_t \in_a^t \tag{3.3}$$

Where, the second term appears due to the attraction, the third term is the randomization parameter, and $\in_a^t$ is a random number vector generated using uniform distribution at time t. If $\gamma = 0$, it reduces to a variant of the particle swarm optimization.

Based on these three rules, the basic steps in FA are explained as follows (Dey et al., 2014):

1. Randomly place a set of N fireflies in the search space within a specific range.
2. Perform watermarking and its corresponding fitness function value for a possible solution for each firefly.
3. Store the solution with the best fitness function as the best scaling factors.
4. Adjust the position of the other fireflies according to the best-fit solution, starting with the firefly having highest light intensity.
5. Update the new firefly position and find the best solution.
6. Repeat steps 2 to 4, N times.

Various image watermarking techniques based on FA are proposed in combination with image processing transforms such as DCT, DWT, SVD, and QR decomposition (Dey et al., 2014; Mishra et al. 2014; Dixit et al., 2016; Swaraja et al., 2016; Guo et al., 2017; Imamoglu et al., 2017). Table 3.7 compares various FA-based digital image watermarking techniques with respect to their fitness functions, embedding domains, and type of watermark detection.

TABLE 3.7 Comparison of FA-Based Digital Image Watermarking Algorithms

Reference Paper	Domain + Techniques Used	Type of Detection	Similarity Measures Used	Fitness Function
Mishra et al. (2014)	Transform (DWT + SVD)	Non-blind	NC	$f = \text{PSNR} + \lambda \cdot \left[\text{NC}(w, w^{*}) + \sum_{i=1}^{N} \text{NC}(w, w_i^{*}) \right]$
Dixit et al. (2016)	Transform (DWT + Shur)	Blind and comparison of coefficients	NC	Not reported
Guo et al. (2017)	Transform (DWT + QR)	Blind and comparison of coefficients	BER and SSIM	$f = [1 - \text{SSIM}(I, \text{WI})] + \lambda \cdot \frac{1}{N} \sum_{i=1}^{N} \text{BER}(w, w_i^{*})$
Swaraja et al. (2016)	Transform (DWT + SVD + Shur)	Blind and comparison of coefficients	NC	Not reported
Imamoglu et al. (2017)	Spatial	Blind	Not reported	Not reported

Where, f, fitness function; NC, normalized cross-correlation; corr, correlation; BER, bit error rate; λ, constant value; N, number of attacks; w, original watermark image; w^*, recovered watermark image; I, original cover image; WI, watermarked image.

FA is mainly used for the calculation of the optimal scaling factor and the selection of the best frequency coefficients of a host image for watermark embedding.

3.5 IMAGE WATERMARKING USING SIMULATED ANNEALING (SA)

Simulated annealing (SA) is a local search optimization algorithm (Aarts et al., 2003; Van Laarhoven et al., 1987) and is based on annealing phenomena, which are thermal processes that find low energy states of an atom in a heat environment. The process contains two steps; (1) increase the temperature of the heat environment to a maximum value at which the atom melts; (2) decrease the temperature of heat environment carefully until the particles arrange themselves into the ground state condition of the atom, which has the minimum energy state of the atom. The value of this state can be obtained only if the maximum temperature is high enough and if the cooling is done slowly. The connection between annealing processes and optimal minimization was established by M. Pincus in 1970 (Pincus, 1970). The annealing process as an optimization technique was proposed by S. Kirkpatrick in 1983 (Kirkpatrick et al., 1983) for a combinational optimization problem. The SA-based optimization process can be performed using the metropolis algorithm (Metropolis et al., 1953) which is based on the Monte Carlo method. The metropolis algorithm generates an optimal solution to combinational optimization problems by assuming an analogy between the input function and many-particle systems with the following assumptions: (1) the solution of the problem is equivalent to the states of a physical system and (2) the value of a solution is equivalent to the "energy" of a state. For the implementation of SA, two functions are required: (1) a successor function that returns a "close" neighboring solution, given the actual optimal value (this function works as a distributive function for the particles of the system) and (2) a target function to optimize, depending on the current state of the system. This function works as the energy of the system. The main advantage of SA is that it avoids being trapped at local minima (Metropolis et al., 1953). The algorithm uses a random search method which accepts both changes in input function f. The probability of optimal value for this algorithm is given as:

$$p = \exp\left(-\frac{\delta f}{T}\right) \tag{3.4}$$

Where, δf is the change in input function and T is a control parameter, which is the analogy by temperature.

The implementation of SA is very easy (Aarts et al., 2003; Van Laarhoven et al., 1987). The input parameters of SA are the possible solutions' values, the generation of random changes in solutions, a mean value of evaluating the problem functions, an initial temperature, and a method or rules for decreasing it in the search process. Only one image watermarking technique is available in the literature (Lin et al., 2010) which used SA for calculation of the optimal scaling factor.

3.6 IMAGE WATERMARKING USING TABU SEARCH

The tabu search (TS) algorithm is based on the process designed to cross boundaries of feasibility or local optimality instead of treating them as barriers (Glover and Laguna, 1998). This algorithm relies on three main strategies (Pham and Karaboga, 2012): (1) a forbidding strategy which controls values in the tabu list, (2) a freeing strategy which controls what the output list of tabu is and when it exists, and (3) a short-term strategy which manages interplay between the other strategies to select best trail solutions. The restriction of tabu is one of the important exceptions in this algorithm. When a tabu has a sufficient evaluation value for attraction where the best optimal result can be achieved, then classification of tabu may be overridden. This condition is called aspiration criterion (Huang et al., 2011).

In this algorithm, first an initial solution is chosen in the tabu list. Then, subsets of solutions are generated such that the solution violates tabu conditions or at least one of the aspiration conditions have to choose a best solution in the subset. If the fitness value of the set is less than fitness value of subset, then tabu and aspiration conditions are updated. If a stopping condition is met, then the algorithm is terminated. Various image watermarking techniques based on TS are proposed in the literature (Latif, 2013; Wang and Niu, 2013; Huang et al., 2011; Huang et al., 2003) based on DCT, Hadamard transform and methods like vector quantization. The TS algorithm is mainly used for the selection of the best frequency coefficients of a host image for watermark embedding. These techniques have less imperceptibility and less robustness against attacks.

REFERENCES

Aarts, E. H. L., J. H. M. Korst, and M. A. Arbib. "Simulated annealing and Boltzmann machines." In *Handbook of Brain Theory and Neural Networks*, (2nd ed) pp. 1039–1044, 2003.

Ali, Musrrat, and Chang Wook Ahn. "An optimal image watermarking approach through cuckoo search algorithm in wavelet domain." *International Journal of System Assurance Engineering and Management* 9(3): 602–611, 2018.

Ali, Musrrat, Chang Wook Ahn, and Millie Pant. "A robust image watermarking technique using SVD and differential evolution in DCT domain." *Optik-International Journal for Light and Electron Optics* 125(1): 428–434, 2014a.

Ali, Musrrat, Chang Wook Ahn, and Millie Pant. "Cuckoo search algorithm for the selection of optimal scaling factors in image watermarking." In *Proceedings of the Third International Conference on Soft Computing for Problem Solving*, pp. 413–425. Springer, New Delhi, 2014b.

Al-Qaheri, Hameed, Abhijit Mustafi, and Soumya Banerjee. "Digital watermarking using ant colony optimization in fractional Fourier domain." *Journal of Information Hiding and Multimedia Signal Processing* 1(3): 179–189, 2010.

Antoniou, Andreas, and Wu-Sheng Lu. *The Optimization Problem.* Springer, US, 2007.

Aslantas, Veysel, A. Latif Dogan, and Serkan Ozturk. "DWT-SVD based image watermarking using particle swarm optimizer." In *IEEE International Conference on Multimedia and Expo*, pp. 241–244. IEEE, 2008.

Aslantas, Veysel. "A singular-value decomposition-based image watermarking using genetic algorithm." *AEU-International Journal of Electronics and Communications* 62(5): 386–394, 2008.

Aslantas, Veysel. "An optimal robust digital image watermarking based on SVD using differential evolution algorithm." *Optics Communications* 282(5): 769–777, 2009.

Baranowski, Terranna M., and Eugene J. LeBoeuf. "Consequence management utilizing optimization." *Journal of Water Resources Planning and Management* 134(4): 386–394, 2008.

Bedi, Punam, Roli Bansal, and Priti Sehgal. "Multimodal biometric authentication using PSO based watermarking." *Procedia Technology* 4: 612–618, 2012.

Bedi, Punam, Roli Bansal, and Priti Sehgal. "Using PSO in a spatial domain-based image hiding scheme with distortion tolerance." *Computers & Electrical Engineering* 39(2): 640–654, 2013.

Binitha, S., and S. Siva Sathya. "A survey of bio-inspired optimization algorithms." *International Journal of Soft Computing and Engineering* 2(2): 137–151, 2012.

Chakraborty, Sayan, Sourav Samanta, Debalina Biswas, Nilanjan Dey, and Sheli Sinha Chaudhuri. "Particle swarm optimization-based parameter optimization technique in medical information hiding." In *IEEE International Conference on Computational Intelligence and Computing Research (ICCIC'13)*, pp. 1–6. IEEE, 2013.

Chatterjee, Sankhadeep, Sarbartha Sarkar, Nilanjan Dey, Amira S. Ashour, and Soumya Sen. "Hybrid non-dominated sorting genetic algorithm: II-neural network approach." In *Advancements in Applied Metaheuristic Computing*, pp. 264–286. IGI Global, 2018.

Chen, Yongchang, Weiyu Yu, and Jiuchao Feng. "A reliable svd-dwt based watermarking scheme with artificial bee colony algorithm." *International Journal of Digital Content Technology and Its Applications* 6(22): 430, 2012.

Chu, Shu-Chuan, Hsiang-Cheh Huang, Yan Shi, Ssu-Yi Wu, and Chin-Shiuh Shieh. "Genetic watermarking for zerotree-based applications." *Circuits, Systems & Signal Processing* 27(2): 171–182, 2008.

Dey, Nilanjan, Amira S. Ashour, Samsad Beagum, Dimitra Sifaki Pistola, Mitko Gospodinov, Evgeniya Peneva Gospodinova, and João Manuel R. S. Tavares. "Parameter optimization for local polynomial approximation-based intersection confidence interval filter using genetic algorithm: an application for brain MRI image de-noising." *Journal of Imaging* 1(1): 60–84, 2015.

Dey, Nilanjan, Sourav Samanta, Sayan Chakraborty, Achintya Das, Sheli Sinha Chaudhuri, and Jasjit S. Suri. "Firefly algorithm for optimization of scaling factors during embedding of manifold medical information: an application in ophthalmology imaging." *Journal of Medical Imaging and Health Informatics* 4(3): 384–394, 2014.

Dey, Nilanjan, Sourav Samanta, Xin-She Yang, Achintya Das, and Sheli Sinha Chaudhuri. "Optimisation of scaling factors in electrocardiogram signal watermarking using cuckoo search." *International Journal of Bio-Inspired Computation* 5(5): 315–326, 2013.

Dixit, Akanksha, Pankaj Sharma, and Vyom Kulshreshtha. "Blind video watermarking based on DWT-SHUR and optimized firefly algorithm." *International Journal of Computer Applications* 147(1): 30–36, 2016.

Dorigo, Marco, Vittorio Maniezzo, and Alberto Colorni. "Ant system: optimization by a colony of cooperating agents." *IEEE Transactions on Systems, Man, and Cybernetics, Part B (Cybernetics)* 26(1): 29–41, 1996.

Fakhari, Pegah, Ehsan Vahedi, and Caro Lucas. "Protecting patient privacy from unauthorized release of medical images using a bio-inspired wavelet-based watermarking approach." *Digital Signal Processing* 21(3): 433–446, 2011.

Farhan, Asma Ahmad, and Sana Bilal. "A novel fast and robust digital image watermarking using Bee Algorithm." In *Multitopic Conference (INMIC), 2011 IEEE 14th International*, pp. 82–86. IEEE, 2011.

Findik, Oğuz, İsmail Babaoğlu, and Erkan Ülker. "A color image watermarking scheme based on hybrid classification method: particle swarm optimization and k-nearest neighbor algorithm." *Optics Communications* 283(24): 4916–4922, 2010.

Ghosh, Tarun Kumar, and Sanjoy Das. "A novel hybrid algorithm based on firefly algorithm and differential evolution for job scheduling in computational grid." *International Journal of Distributed Systems and Technologies (IJDST)* 9(2): 1–15, 2018.

Glover, Fred. "Future paths for integer programming and links to artificial intelligence." *Computers & Operations Research* 13(5): 533–549, 1986.

Glover, Fred. "Heuristics for integer programming using surrogate constraints." *Decision Sciences* 8(1): 156–166, 1977.

Glover, Fred, and Manuel Laguna. "Tabu search." In *Handbook of Combinatorial Optimization*, pp. 2093–2229. Springer, Boston, MA, 1998.

Golshan, F., and K. Mohammadi. "A hybrid intelligent SVD-based perceptual shaping of a digital image watermark in DCT and DWT domain." *The Imaging Science Journal* 61(1): 35–46, 2013.

Guo, Yong, Bing-Zhao Li, and Navdeep Goel. "Optimised blind image watermarking method based on firefly algorithm in DWT-QR transform domain." *IET Image Processing* 11(6): 406–415, 2017.

Holland, John. *Adaptation in Natural and Artificial Systems: An Introductory Analysis with Applications to Biology, Control and Artificial Intelligence.* MIT Press, Cambridge, MA, 1992.

Huang, Hsiang-Cheh, Jeng-Shyang Pan, Feng-Hsing Wang, and Chin-Shiuh Shieh. "Robust image watermarking with tabu search approaches." In *IEEE Int. Symposium on Consumer Electronics (ISCE 2003)*, Sydney, Australia, 2003.

Huang, Hsiang-Cheh, Jeng-Shyang Pan, Yu-Hsiu Huang, Feng-Hsing Wang, and Kuang-Chih Huang. "Progressive watermarking techniques using genetic algorithms." *Circuits, Systems & Signal Processing* 26 (5): 671–687, 2007.

Huang, Hsiang-Cheh, Shu-Chuan Chu, Jeng-Shyang Pan, Chun-Yen Huang, and Bin-Yih Liao. "Tabu search based multi-watermarks embedding algorithm with multiple description coding." *Information Sciences* 181(16): 3379–3396, 2011.

Imamoglu, Mustafa Bilgehan, Mustafa Ulutas, and Guzin Ulutas. "A new reversible database watermarking approach with firefly optimization algorithm." *Mathematical Problems in Engineering* 2017: 1–14, 2017.

Issa, Mohamed. "Digital image watermarking performance improvement using bio-inspired algorithms." In *Advances in Soft Computing and Machine Learning in Image Processing*, pp. 683–698. Springer, Cham, 2018.

Karaboga, Dervis. *An Idea Based on Honey Bee Swarm for Numerical Optimization*, vol. 200. Technical report-tr06, Erciyes University, Engineering Faculty, Computer Engineering Department, 2005.

Kennedy, R. J. and Eberhart, "Particle swarm optimization." In *Proceedings of IEEE International Conference on Neural Networks IV, pages*, vol. 1000. 1995.

Kirkpatrick, Scott, C. Daniel Gelatt, and Mario P. Vecchi. "Optimization by simulated annealing." *Science* 220 (4598): 671–680, 1983.

Koza, John R. *Genetic Programming II, Automatic Discovery of Reusable Subprograms.* MIT Press, Cambridge, MA, 1992.

Koza, John R. "Genetic programming as a means for programming computers by natural selection." *Statistics and Computing* 4 (2): 87–112, 1994.

Latif, Ali Mohammad. "An adaptive digital image watermarking scheme using fuzzy logic and tabu search." *Journal of Information Hiding and Multimedia Signal Processing* 4(4): 250–271, 2013.

Lee, Jiann-Shu, Jing-Wein Wang, and Kung-Yo Giang. "A new image watermarking scheme using multi-objective bees algorithm." *Applied Mathematics & Information Sciences* 8(6): 2945–2953, 2014.

Li, Jianzhong. "An optimized watermarking scheme using an encrypted gyrator transform computer generated hologram based on particle swarm optimization." *Optics Express* 22(8): 10002–10016, 2014.

Li, Xiaoxia, and Jianjun Wang. "A steganographic method based upon JPEG and particle swarm optimization algorithm." *Information Sciences* 177(15): 3099–3109, 2007.

Lin, Guo-Shiang, Yi-Ting Chang, and Wen-Nung Lie. "A framework of enhancing image steganography with picture quality optimization and anti-steganalysis based on simulated annealing algorithm." *IEEE Transactions on Multimedia* 12(5): 345–357, 2010.

Loukhaoukha, Khaled, Jean-Yves Chouinard, and Mohamed Haj Taieb. "Optimal image watermarking algorithm based on LWT-SVD via multi-objective ant colony optimization." *Journal of Information Hiding and Multimedia Signal Processing* 2(4): 303–319, 2011.

Maity, Santi P., and Malay K. Kundu. "Genetic algorithms for optimality of data hiding in digital images." *Soft Computing* 13(4): 361–373, 2009.

Maity, Santi P., Seba Maity, Jaya Sil, and Claude Delpha. "Collusion resilient spread spectrum watermarking in M-band wavelets using GA-fuzzy hybridization." *Journal of Systems and Software* 86(1): 47–59, 2013.

Memon, Mudasir, Mekhilef Saad, and Marizan Mubin. "Selective harmonic elimination in multilevel inverter using hybrid APSO algorithm." *IET Power Electronics* 11(10): 1673–1680, 2018.

Metropolis, Nicholas, Arianna W. Rosenbluth, Marshall N. Rosenbluth, Augusta H. Teller, and Edward Teller. "Equation of state calculations by fast computing machines." *The Journal of Chemical Physics* 21(6): 1087–1092, 1953.

Mishra, A., and C. Agarwal. "Toward optimal watermarking of grayscale images using the multiple scaling factor–based cuckoo search technique." In *Bio-Inspired Computation and Applications in Image Processing*, pp. 131–155. Academic Press, London, 2016.

Mishra, Anurag, Charu Agarwal, Arpita Sharma, and Punam Bedi. "Optimized gray-scale image watermarking using DWT–SVD and Firefly Algorithm." *Expert Systems with Applications* 41(17): 7858–7867, 2014.

Nelder, J. A., and Mead, R. A simplex method for function minimization. *The Computer Journal* 7(4): 308–313, 1965.

Nurdin, Hardisal, Muhammad Zarlis, and Erna Budhiarti Nababan. "Adaptive watermarking technique using micro genetic algorithm." *Journal of Inotera* 1(1): 64–70, 2017.

Peng, Hong, Jun Wang, Mario J. Pérez-Jiménez, and Agustín Riscos-Núñez. "The framework of P systems applied to solve optimal watermarking problem." *Signal Processing* 101: 256–265, 2014.

Pham, Duc, and Dervis Karaboga. *Intelligent Optimisation Techniques: Genetic Algorithms, Tabu Search, Simulated Annealing and Neural Networks*. Springer Science & Business Media, 2012.

Pincus, Martin. "Letter to the editor—a Monte Carlo method for the approximate solution of certain types of constrained optimization problems." *Operations Research* 18(6): 1225–1228, 1970.

Powell, M. J. D. "A survey of numerical methods for unconstrained optimization." *SIAM Review* 12(1): 79–97, 1970.

Rao, R. Surya Prakasa, and P. Rajesh Kumar. "A novel signature based watermarking approach for color images based on GA and PSO." *International Journal of Computational Engineering Research (IJCER)* 7(7): 36–46, 2017.

Rohani, Mohsen, and Alireza Nasiri Avanaki. "A watermarking method based on optimizing SSIM index by using PSO in DCT domain." In *14th International CSI Computer Conference*, Tehran, Iran, 2009.

Run, Ray-Shine, Shi-Jinn Horng, Jui-Lin Lai, Tzong-Wang Kao, and Rong-Jian Chen. "An improved SVD-based watermarking technique for copyright protection." *Expert Systems with Applications* 39(1): 673–689, 2012.

Sanku, Deepika, Sampath Kiran, Tamirat Tagesse Takore, and P. Rajesh Kumar. "Digital Image Watermarking in RGB Host Using DWT, SVD, and PSO Techniques." In *Proceedings of 2nd International Conference on Micro-Electronics, Electromagnetics and Telecommunications*, pp. 333–342. Springer, Singapore, 2018.

Shieh, Chin-Shiuh, Hsiang-Cheh Huang, Feng-Hsing Wang, and Jeng-Shyang Pan. "Genetic watermarking based on transform-domain techniques." *Pattern Recognition* 37(3): 555–565, 2004.

Storn, Rainer, and Kenneth Price. "Differential evolution–a simple and efficient heuristic for global optimization over continuous spaces." *Journal of Global Optimization* 11, (4): 341–359, 1997.

Swaraja, K., Y. Madhaveelatha, and V. S. K. Reddy. "Robust video watermarking by amalgamation of image transforms and optimized firefly algorithm." *International Journal of Applied Engineering Research* 11(1): 216–225, 2016.

Thakkar, Falgun, and Vinay Kumar Srivastava. "A particle swarm optimization and block-SVD-based watermarking for digital images." *Turkish Journal of Electrical Engineering & Computer Sciences* 25(4): 3273–3288, 2017.

Van Laarhoven, Peter J. M., and Emile H. L. Aarts. "Simulated annealing." In *Simulated Annealing: Theory and Applications*, pp. 7–15. Springer, Dordrecht, 1987.

Vellasques, Eduardo, Robert Sabourin, and Eric Granger. "A high throughput system for intelligent watermarking of bi-tonal images." *Applied Soft Computing* 11(8): 5215–5229, 2011.

Vellasques, Eduardo, Robert Sabourin, and Eric Granger. "Fast intelligent watermarking of heterogeneous image streams through mixture modeling of PSO populations." *Applied Soft Computing* 13(6): 3130–3148, 2013.

Venter, Gerhard. "Review of optimization techniques." *Encyclopedia of Aerospace Engineering* 1: 1–12, 2010.

Verma, Vibha, Vinay Kumar Srivastava, and Falgun Thakkar. "DWT-SVD based digital image watermarking using swarm intelligence." In *International Conference on Electrical, Electronics, and Optimization Techniques (ICEEOT'16)*, pp. 3198–3203. IEEE, 2016.

Waleed, Jumana, Huang Dong Jun, Saad Hameed, and May Kamil. "Optimal positions selection for watermark inclusion based on a nature inspired algorithm." *International Journal of Signal Processing, Image Processing and Pattern Recognition* 8(1): 147–160, 2015.

Waleed, Jumana, Huang Dong Jun, Thekra Abbas, Saad Hameed, and Hiyam Hatem. "A survey of digital image watermarking optimization based on nature inspired algorithms NIAs." *International Journal of Security and Its Applications* 8(6): 315–334, 2014.

Wang, Shen, and Xiamu Niu. "Hiding traces of double compression in JPEG images based on Tabu Search." *Neural Computing and Applications* 22(1): 283–291, 2013.

Wang, Yuh-Rau, Wei-Hung Lin, and Ling Yang. "An intelligent watermarking method based on particle swarm optimization." *Expert Systems with Applications* 38(7): 8024–8029, 2011.

Wu, Chun-Ho, Y. Zheng, W. H. Ip, C. Y. Chan, K. L. Yung, and Z. M. Lu. "A flexible H. 264/AVC compressed video watermarking scheme using particle swarm optimization-based dither modulation." *AEU-International Journal of Electronics and Communications* 65(1): 27–36, 2011.

Yang, Xin-She, and Suash Deb. "Cuckoo search via Lévy flights." In *World Congress on Nature & Biologically Inspired Computing, NaBIC 2009*, pp. 210–214. IEEE, 2009.

Yang, Xin-She, Seyyed Soheil Sadat Hosseini, and Amir Hossein Gandomi. "Firefly algorithm for solving non-convex economic dispatch problems with valve loading effect." *Applied Soft Computing* 12(3): 1180–1186, 2012.

Yang, Xin-She. "Engineering optimizations via nature-inspired virtual bee algorithms." In *International Work-Conference on the Interplay Between Natural and Artificial Computation*, pp. 317–323. Springer, Berlin, Heidelberg, 2005.

CHAPTER 4

Hardware-Based Implementation of Watermarking

SOFTWARE-BASED IMPLEMENTATION OF WATERMARKING is easy to use and flexible, as it can be easily modified and upgraded. Most of the efficient and robust watermarking algorithms embed watermarks in transform domains and include forward and inverse transform operations, which are complex and computationally expensive. However, their speed is limited, and they are vulnerable to offline attacks. The speed of these algorithms depends on the processor. Implementing real-time image watermarking algorithms on serial processors involves lots of computations due to their high resolutions (AlAli et al., 2013), and they have some additional constraints related to peripheral devices and memory. Hence, there is a requirement for dedicated and customized hardware to support image watermarking in real time. The advantages of customized hardware realization of watermarking include:

- Low cost
- Lower power consumption
- Less execution time
- Reduced chip area
- Reliability
- Supports real-time processing
- Supports stand-alone implementation without using a personal computer (PC)

This chapter discusses the hardware design technologies that enable digital image watermarking *for Digital Signal Processors* (DSPs), *Field Programmable Gate Arrays* (FPGAs), *and Application-Specific Integrated Circuits* (ASIC) *chips* along with the hardware-software co-design in Xilinx System Generator (XSG). Various hardware watermarking implementation approaches that use FPGAs, DSPs, and ASIC boards in spatial, transform, and hybrid domains are compared. This comparison is done in terms of the board and platform used, clock frequency, power consumption, maximum peak signal to noise ratio (PSNR) (dB) achieved, and the type of host image support provided.

4.1 INTRODUCTION

There are two types of hardware design technologies: (1) ASIC, which are full custom, and (2) Programmable/reconfigurable devices such as DSPs and FPGAs, which are semi-custom. The parameters to be considered while selecting the design technology are cost, complexity, flexibility, design time, reconfigurability, performance, and application. ASIC offers the highest performance due to pipelining, concurrency, parallelism, and clubbing of single clock operations. On the flip side, ASICs are not flexible and reconfigurable. Furthermore, a single error in the fabrication process makes the chip useless. DSPs perform extremely well when processing complex images. The design complexity and performance of DSP processors are between those of a PC and ASIC. The DSPs may be programmed using their own assembly language, or in C using associated C compilers.

FPGAs are reconfigurable devices that exhibit the benefits of general-purpose processors while demonstrating performance levels that are close to ASICs. The FPGAs are reliable, use low power, run at low clock frequency, support parallelism and pipelining techniques, simplifies debugging and verification, minimizes time to market, allows rapid prototyping and design reuse, supports upgradations, and performs reconfiguration at low cost and in less time. Parallel execution of multiple Multiply and Accumulate (MAC) operations is possible with FPGA. Because of these functional advantages, and due to improvements in architectures and size, the FPGAs have become widely used target devices for image and video processing applications (AlAli et al., 2013).

4.2 HARDWARE-BASED IMPLEMENTATION OF DIGITAL IMAGE WATERMARKING

Development of real-time watermarking includes two steps: (1) algorithm development and simulation and (2) hardware implementation and testing. In the process of designing hardware systems using reconfigurable devices, the first step is to write a program using hardware description language (HDL). The HDL codes represent synthesizable register-transfer level (RTL) models of the design and can be simulated by tools such as ModelSim. The two principal languages used for configuring FPGAs are Very High Speed Integrated Circuits Hardware Description Language (VHDL) and Verilog HDL (Verilog). Many C-based hardware descriptive languages, such as System C, have been developed. However, the synthesis processes of such languages are not automated and include human interaction. To overcome this problem, Handel-C and the associated compilers are developed by Celoxia Ltd. for direct hardware implementation from C-based language description.

The system level hardware programming languages are hardware specific and demand the designers to have enough knowledge of hardware and its related implementation. Hence, there is a need to translate high-level languages into hardware descriptions. For example, if the program is written in MATLAB®, then behavioral synthesis tools, such as MATCH compilers, are used to automatically generate synthesizable RTL models in VHDL/Verilog. The hardware implementation of watermarking using FPGA/DSP boards, using behavioral modeling of XSG architecture and MATLAB Simulink, is a very common practice in image watermarking. In developing the watermark Simulink model, the Video and Image Processing Blockset along with the signal processing tool box of MATLAB are widely used.

In the last two decades, many simulation and hardware implementations of watermarking have been developed (Joshi et al., 2012; Lakshmi et al., 2017) based on different domains: Discrete Fourier Transform (DFT), Discrete Cosine Transform (DCT), and Discrete Wavelet Transform (DWT) (Kothari et al., 2018). While the hardware-based implementations of watermarking can be done by using both DSP boards and FPGA, it is preferred to use FPGA in general as they require less power and area compared to DSP boards. Also, ASIC and FPGA are easily compatible with simulation programs in which coding and watermarking are done smoothly. On the other hand, DSPs have high storage memory compared to the other two boards. Thus, DSPs are better suited for watermarking big multimedia files, such as video. Many hardware-based implementations of watermarking systems use Xilinx Integrated Synthesis Environment (ISE) and ModelSim in combination with MATLAB to synthesize and process the HDL algorithms on FPGA, as they are less expensive in time and cost. Designers can design and simulate a model-based system using Simulink/Xilinx library. A HDL Coder tool automatically generates synthesizable HDL code mapped to Xilinx pre-optimized algorithms.

4.2.1 Hardware-Based Implementation of Watermarking Using DSP Boards

Various DSPs, such as TMS320C5410, TMS320C670, etc. are used for hardware-based implementations of watermarking. The choice of these processors depends on the type of watermarking technique employed, type of multimedia files involved, and the processing time. The basic hardware-based implementation flow for watermarking using a DSP platform is shown in Figure 4.1. In this flow, the application software, such as MATLAB simulation tool, Visual CV, etc. is used for implementing watermarking algorithms on a PC. The code of the algorithm is transferred to the DSP board using a Host-Port Interface (HPI) bus and Industry Standard Architecture (ISA) bus. The synchronization signals are used to synchronize the PC and DSP board. The DSP board can operate in two modes: (1) Offline data generation and (2) Live data generation. In both modes, the host-PC reads and writes multimedia data directly into the internal memory of a DSP board using the HPI cable.

- In offline data generation, the multimedia file from the host-PC is transferred to the DSP board, to watermark the multimedia files using codes on the DSP board. This data can be sent back to the host-PC where it can be displayed.

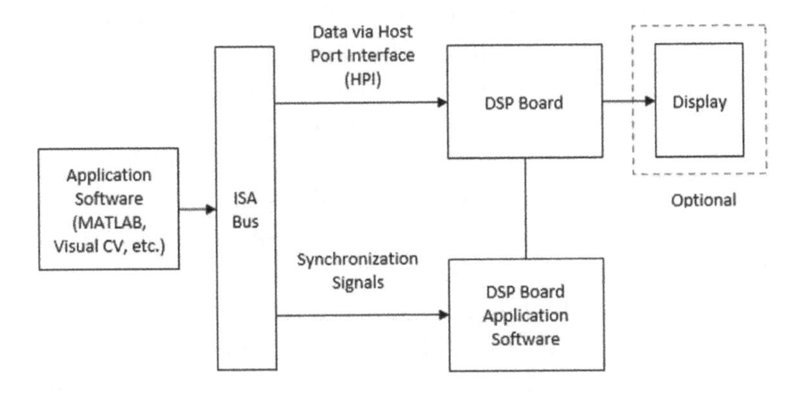

FIGURE 4.1 Implementation of watermarking using DSP platform.

- In live data generation, the multimedia file from the host-PC is transferred to the DSP board and this data can be directly displayed on a monitor or projector using the Video Graphics Array (VGA) mode of the DSP board. The monitor or projector can be connected using a High-Definition Multimedia Interface (HDMI) cable. In this mode, synchronization between PC and DSP board is required.

4.2.2 Hardware-Based Implementation of Watermarking Using FPGA/ASIC Chip

A FPGA design generally contains a large number of simple processors for parallel processing (Khan et al., 2015). The FPGA is an integrated circuit (IC) composed of flip flops, Random Access Memory (RAM), programmable switches and interconnecting lines, and a large number of programmable logic blocks which are often called as Look up Tables (LUTs) (Pemmaraju et al., 2017). Programming an FPGA includes specifying the logic function of each logic cell and the interconnections. Low-power Very Large Scale Integration (VLSI) features, such as clock gating, dynamic clocking, and multiple supply voltages, can be used (Ghosh et al., 2014). While GPUs and DSPs are designed for a general set of functions, FPGA provides reconfigurable solutions and offer higher throughputs and data rates over DSP processors.

For an application that involves multiple operations which can be implemented in parallel, performance gain can be achieved with FPGAs as they exploit inherent temporal and spatial parallelism. An application must be designed to maximize the utilization of the FPGA on chip resources, such as on-board memories (SRAM, SDRAM, etc.), DSP units, and Block RAM, to maximally exploit the inherent parallelism of an FPGA. Note that simply porting an algorithm onto an FPGA is inefficient as most image processing algorithms have already been optimized for execution on a serial processor. Though FPGAs provide throughput advantages over DSPs, for applications that demand high accuracy and floating-point operations, floating-point DSPs may be chosen for better chip area advantages over FPGAs.

FPGA boards and ASIC chips, such as Virtex, Cyclone, Altera, Xilinx, and Customized IC with various μm CMOS technology, etc., are used for hardware implementation. The choice of these boards or chips depends on the type of watermarking technique, application, and multimedia files that are being used, as well as processing time requirements. As a common practice, hardware-based implementations of watermarking are designed and tested on a FPGA board before the watermarking chip is made on an ASIC. In the designing process, various VLSI simulation tools such as Quartus II, Xilinx ISE design, etc. are used for coding of watermark embedding and extraction algorithms. These designs also use the MATLAB tool to read and write the multimedia files and to store the generated watermarked files on the host-PC. Two simulation tools are used to design watermarking algorithms:(1) VLSI tools and (2) DSP tools. The DSP simulation tool reads the multimedia file, converts it into the digital format, and feeds it to the VLSI simulation tool, where the watermark embedding and extraction algorithms are designed and implemented, using either VHDL or Verilog. These codes are then converted into a hex file which can be loaded onto a FPGA board to test the algorithms. After successful testing, a CMOS schematic circuit of an algorithm can be generated based on the RTL view, using CMOS layout designing tools, and this is again verified for its performance. Finally, the watermarking IC is built based on the designed CMOS layout. The limitation of this design flow is that once the IC is built no one can change any parameters of the IC. The basic structure of watermarking chips is shown in Figure 4.2. The design flow for watermarking using ASIC is shown in Figure 4.3.

4.2.3 Hardware–Software Co-Simulation

The XSG (Lin et al., 2009) is a useful tool for the development and implementation of real-time image/video/computer vision-based algorithms. Implementation of such algorithms on FPGAs is the

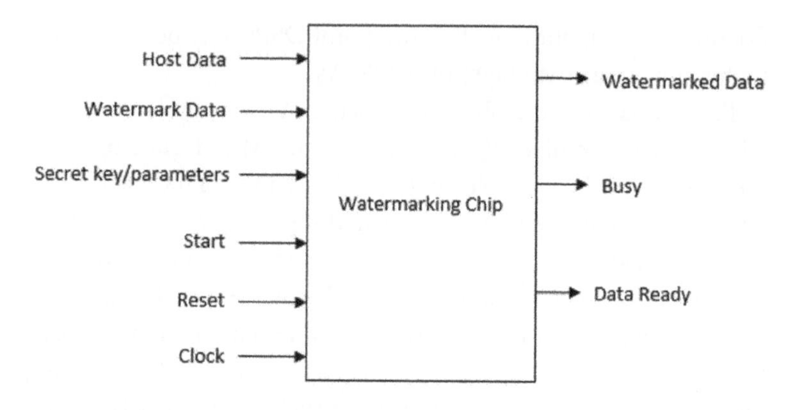

FIGURE 4.2 Basic pins of watermarking chip.

best choice for prototyping, as they are composed of large memories and embedded multipliers, which can process the binary and fixed-point operations at a very high rate, exploiting spatial and temporal parallelism. This section presents the design and implementation flow of hardware–software co-simulation processes using prototype tools such as XSG and MATLAB-Simulink (Madanayake and Len, 2008).

The XSG uses Simulink Blockset for several built-in hardware operations that could be implemented on various Xilinx FPGAs. The MATLAB Simulink is used to program and simulate the model. High-level languages and compilers which are capable of automatically extracting the parallelism from the code are not directly compatible with hardware. The XSG automatically generates a HDL test bench from Simulink models and supports design verification and implementation on DSP/FPGA platforms (Karthigaikumar and Baskaran, 2012).

HDL coders are used to generate VHDL/Verilog code from Simulink models/MATLAB code. ModelSim is used to test the RTL code. The XSG has an integrated design flow to transfer the configuration file that is essential for programming into the FPGA by using hardware co-simulation tools. The blocks of XSG only process fixed-point/Boolean values in contrast to the Simulink blocks,

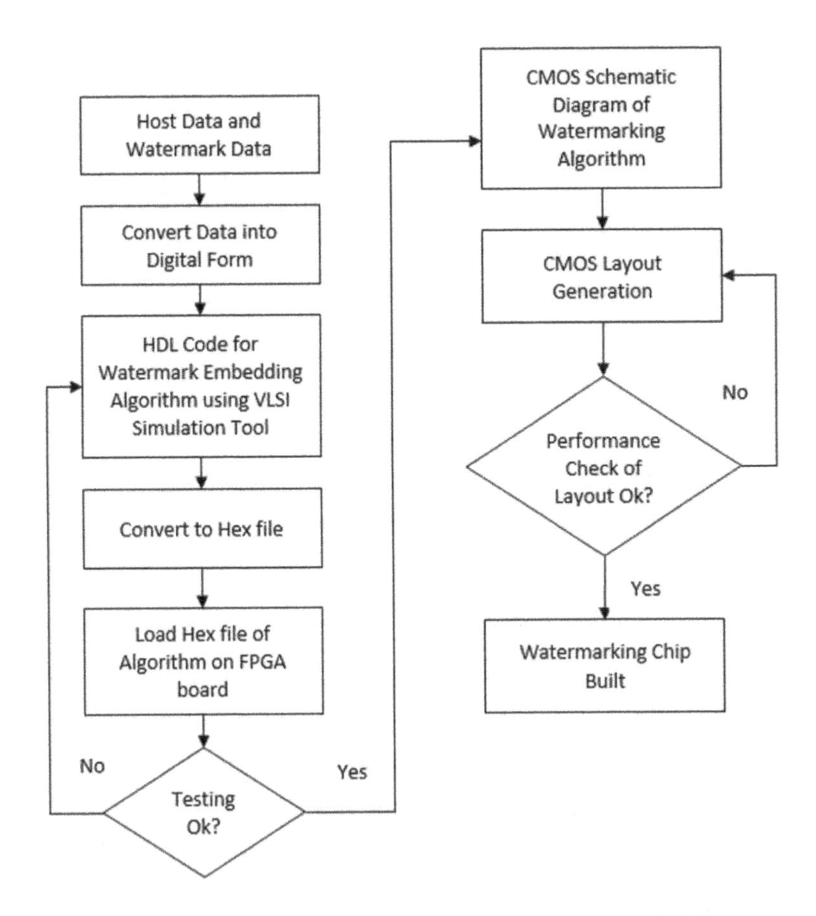

FIGURE 4.3 Watermarking chip generation using ASIC technology.

which can process double-precision floating-point numbers. The gateway blocks are often used for interconnecting Simulink blocks and XSG blocks. The built-in synthesizer of XSG allows the conversion of HDL code into a gate-level netlist and generates a synthesis report, which indicates run time, area, and power (Rahimunnisa et al., 2012). The FPGA configuration files can then be generated by a compilation script and hardware co-simulation blocks (Que et al., 2010). The bit streams are then transferred from the PC to a FPGA board by using a Universal Asynchronous Receiver/Transmitter (UART) serial communication, a JTAG platform, and

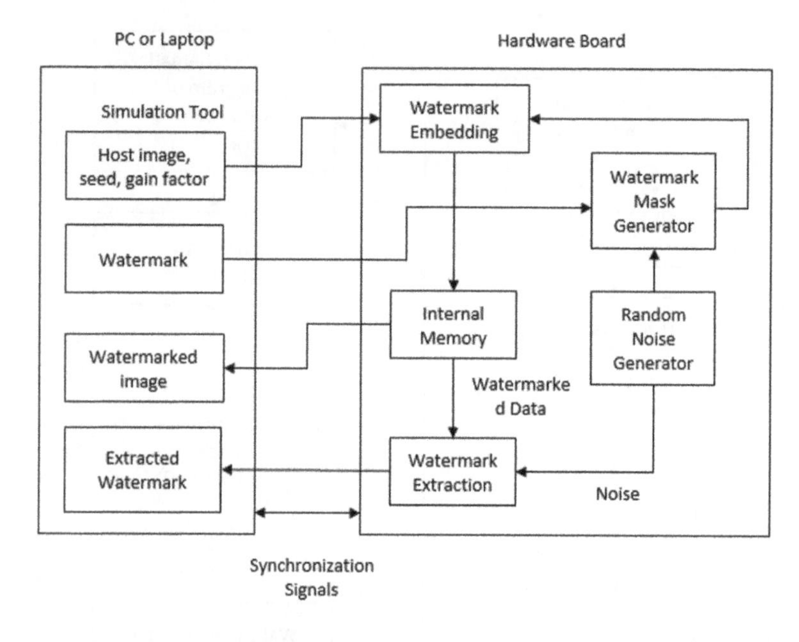

FIGURE 4.4 Simple hardware implementation block diagram of digital watermarking technique.

Universal Serial Bus (USB) cable. After specified processing, the result can be transferred back to a PC for display and validation. Finally, resource utilization and timing reports can be viewed. The target devices can be Spartan/Virtex/Altera boards (Rahimunnisa et al., 2012). The block diagram of simple hardware implementation for digital watermarking is shown in Figure 4.4.

4.3 PERFORMANCE OF HARDWARE-BASED IMPLEMENTATION

In this section, various hardware watermarking implementation approaches using FPGA, DSP, and ASIC boards in spatial, transform, and hybrid domains are compared. There are several approaches for watermarking in the spatial domain. These watermarking techniques may be by the Least Significant Bit (LSB) substitution, additive watermarking, Spread Spectrum (SS)-based watermarking, and reversible watermarking technique.

The hardware implementation of the LSB substitution technique includes a simulation on a PC, using simulation tools and watermark embedding and extraction on a hardware board. The simulation tool is used for reading, storing, and displaying images. The hardware board performs watermark insertion into the host image and outputs watermarked images for display or storage on a PC. Implementation of these techniques using various hardware boards and technology is available (Vinh and Koo, 2017; Shet et al., 2017; Rajagopalan et al., 2014; Samanta et al., 2008; Nelson et al., 2005; Garimella et al., 2004; Garimella et al., 2003), and they are compared in Tables 4.1 through 4.8, in terms of board and platform used, clock frequency, power consumption, maximum PSNR (dB) achieved, and type of host image supported.

Additive watermarking approaches exploit the usage of the Pseudo-Random Noise (PN) sequences when embedding the scaled watermarks in an additive fashion using Linear Feedback Shift Registers (LFSR) to the host image (Shih, 2017). These sequences are preferred as they can affect the pixel values in an imperceptible way due to their low magnitudes. These random sequences are generated using a seed, which acts as a secret key. The additive watermarking techniques are good at imperceptibility and robustness. If the strength of noise increases, the visual quality of the extracted watermark decreases. Furthermore, the payload capacity of this technique is less and cannot obtain satisfactory results for color watermarks.

In additive watermarking techniques, the watermark blocks are selected sequentially for hiding watermark bits, and hence, they are not secure from attacks. In SS-based watermarking techniques, the watermark bits are randomly scattered throughout the host image in order to increase the payload capacity and to improve their robustness against attacks. SS techniques are secure, imperceptible, and robust. However, the visual quality of watermarked data degrades when the gain factor increases.

Reversible watermarking techniques provides a solution for the extraction of host data without a loss besides the authenticating

TABLE 4.1 LSB Substitution-Based Techniques

Reference Paper	HDL/ Compiler	Hardware Board	Hardware Platform	Clock Frequency	Power Consumption	Maximum PSNR(dB)	Type of Host Image	Visible/ Invisible	Robust/ Fragile
Vinh and Koo (2017)	Verilog HDL	Altera DE2 Kit	FPGA	50 MHz	N/A	Around 52	Color Images	Invisible	Robust
Shet et al. (2016)	Verilog HDL	Xilinx Virtex-II Pro XC2V500FG256-6	FPGA	50 MHz	N/A	84.46	Grayscale Images	Invisible	Robust
Rajagopalan et al. (2014)	Not Reported	Cyclone II EP2C35F672C6	FPGA	50 MHz	N/A	60.83	Color Images	Invisible	Not reported
Samanta et al. (2008)	VHDL	Single Electron Tunneling (SET) and Xilinx ISE 9.1i	FPGA	N/A	N/A	42.11	Grayscale Images	Invisible	Robust
Nelson et al. (2005)	MATLAB	Chip with 0.18 μm CMOS 6-metal Technology	ASIC	N/A	N/A	Not reported	Grayscale Images	Invisible	Fragile
Garimella et al. (2004)	VHDL	Chip with 0.13 μm CMOS 6-metal Technology	ASIC	166.6 MHz	9.1941 mW	N/A	Color Images	Invisible	Fragile
Garimella et al. (2003)	VHDL	Chip with 0.13 μm CMOS 6-metal Technology	ASIC	100 MHz	37.6 μW	64.0399	Grayscale Images	Invisible	Fragile

TABLE 4.2 Additive Watermarking

Reference Paper	HDL/ Compiler	Hardware Board	Hardware Platform	Clock Frequency	Power Consumption	Maximum PSNR (dB)	Type of Host Image	Visible/ Invisible	Robust/ Fragile
Mohanty et al. (2005)	VHDL	Chip with 0.35 μm CMOS 6-metal Technology	ASIC	151 MHz	6.93 mW	25	Grayscale images	Visible	Robust
Mohanty and Nayak (2004)	VHDL	Synplify ProTM Tool and XCV50-BG256-6 Xilinx Virtex-II	FPGA	50 MHz	N/A	N/A	Grayscale images	Invisible	Robust
Petitjean et al. (2002)	C language	FPGA Board and DSP Board	FPGA and DSP	50 MHz and 250 MHz	N/A	38.69	Grayscale videos	Invisible	Robust
Maes et al. (2000)	Not reported	Custom IC	ASIC	Not reported	N/A	N/A	Not reported	Invisible	Robust

TABLE 4.3 Spread Spectrum Watermarking Techniques

Reference Paper	HDL/ Compiler	Hardware Board	Hardware Platform	Clock Frequency	Power Consumption	Maximum PSNR (dB)	Type of Host Image	Visible/ Invisible	Robust/ Fragile
Ghosh et al. (2012)	MATLAB and VHDL	Xilinx Virtex and Virtex-II Pro XC2VP30	FPGA	82.26 MHz	1300 mW	36.12	Grayscale images	Invisible	Robust
Tsai and Wu (2003)	Not reported	Custom IC	ASIC	N/A	N/A	N/A	Videos	Invisible	Robust

TABLE 4.4 Reversible Watermarking Technique

Reference Paper	HDL/ Compiler	Hardware Board	Hardware Platform	Clock Frequency	Power Consumption	Maximum PSNR (dB)	Type of Host Image	Visible/ Invisible	Robust/ Fragile
Lakshmi et al. (2017)	Verilog HDL	Vivado Tool and Zynq-7000 Xilinx kit	FPGA	N/A	6.4 W	N/A	Grayscale images	Invisible	Robust
Maity and Maity (2017)	VHDL	XC4VLX80-12FF1148	FPGA	95.3 MHz	636 mW	44.3	Grayscale images	Invisible	Fragile
Maity and Maity (2014)	VHDL	Xilinx Spartan, XC3S1600E	FPGA	98.76 MHz	750 mW	N/A	Grayscale images	Invisible	Robust
Zavaleta et al. (2008)	VHDL	Xilinx ISE 9.2i Tools and Spartan 3 XC3s500e-50fg320	FPGA	N/A	N/A	N/A	Grayscale medical images	Invisible	Not reported

TABLE 4.5 DCT-Based Watermarking

Reference Paper	HDL/ Compiler	Hardware Board	Hardware Platform	Clock Frequency	Power Consumption	Maximum PSNR (dB)	Type of Host Image	Visible/ Invisible	Robust/ Fragile
Tsai and Yang (2017)	C/C++ compiler	TMS320C6701	DSP	167 MHz	N/A	N/A	Grayscale images	Invisible	Robust
Shah (2017)	Android	Android Smart Phone	Android	N/A	N/A	N/A	Color images	Invisible	Robust
Roy et al. (2013)	Verilog HDL	Mentor's ModelSim Tool + Altera Cyclone EPIC20	FPGA + ASIC	40 MHz	270 mW	44	Grayscale videos	Invisible	Semi fragile
Morita et al. (2009)	MATLAB and Hyper Terminal	Xilinx Virtex-II Pro Board	FPGA	131.092 MHz	Not reported	N/A	Color images	Visible	Robust
Mohanty et al. (2007)	VHDL	Synplify ProTM Tool and XCV50-BG256-6 Xilinx Virtex-II	FPGA	151 MHz	24 mW	50	Grayscale images	Invisible	Robust/ fragile
Mohanty et al. (2003)	VHDL	Virtuoso Layout Tool	FPGA and ASIC	151 MHz	24 mW	N/A	Grayscale images	Invisible	Robust/ fragile
Lim et al. (2003)	VHDL	EP10K1000ARC240-3	Camera	50 MHz	N/A	N/A	Color images	Visible	Robust
Tsai and Lu (2001)	VHDL	Chip with 0.35 μm CMOS Technology	ASIC	151 MHz	107.6 μW	N/A	Grayscale images	Invisible	Robust

TABLE 4.6 DWT-Based Watermarking

Reference Paper	HDL/ Compiler	Hardware Board	Hardware Platform	Clock Frequency	Power Consumption	Maximum PSNR (dB)	Type of Host Image	Visible/ Invisible	Robust/ Fragile
Mulani and Mane (2017)	Verilog HDL and MATLAB	Xilinx ISE Design Suite 13.1 and xc6vcx75t-2ff484 Xilinx kit	FPGA	228.064 MHz	N/A	N/A	Grayscale images	Invisible	Robust
Lad et al. (2011)	VHDL and MATLAB	Xilinx Chip-Scope Pro tool and Virtex-II Pro XC2VP30	FPGA	29.107 MHz	3.75 mW	44.408	Grayscale images	Not reported	Robust
Mathai et al. (2003)	Not reported	Chip with 0.18 μm CMOS Technology	ASIC	75 ` MHz	160 mW	40	Grayscale images	Invisible	Robust
Hsiao et al. (2001)	Not reported	Custom IC	ASIC	Not Reported	N/A	N/A	Grayscale Images	Invisible	Robust

TABLE 4.7 Advanced Watermarking Technique

Reference Paper	HDL/ Compiler	Hardware Board	Hardware Platform	Clock Frequency	Power Consumption	Maximum PSNR (dB)	Type of Host Image	Visible/ Invisible	Robust/ Fragile
Nayak et al. (2017)	VHDL	Xilinx ISE 14.3 Simulation Tools, Kintex 7 Kit (7k325tfbg676-3) and Spartan 6 (651x45tfgg484-3)	FPGA	351.457 MHz	5.029 mW	45.23	Grayscale images	Invisible	Robust
Maity and Kundu (2013)	VHDL	Xilinx Spartan, XCS05	FPGA	80 MHz	N/A	41.02	Grayscale images	Invisible	Not reported

TABLE 4.8 Hybrid Watermarking

Reference Paper	HDL/ Compiler	Hardware Board	Hardware Platform	Clock Frequency	Power Consumption	Maximum PSNR (dB)	Type of Host Image	Visible/ Invisible	Robust/ Fragile
Harini et al. (2017)	C/C++ compiler and GNU compiler	Xilinx Platform Studio (XPS)	FPGA	80.749 MHz	N/A	N/A	Color images	Invisible	Fragile
Venugopala et al. (2017)	Android	Android Smart Phone	Android	N/A	669 mW	N/A	Color videos	Invisible	Fragile
Joshi et al. (2012)	MATLAB and VHDL	Xilinx project navigator ISE 9.1 and SPARTAN 3E Xc3s500e-4fg320	FPGA	N/A	69 μW	N/A	Color images and color videos	Invisible	Robust
Karmani et al. (2009)	VHDL	Altera Stratix II EP2S606C57ES	FPGA	100 MHz	N/A	31 to 32	Grayscale images	Invisible	Fragile
Maity et al. (2009)	VHDL	Xilinx Spartan, XCS40L	FPGA	80 MHz	0.3 mW	Not reported	Grayscale images	Visible	robust
Maity et al. (2007)	VHDL	Xilinx Spartan, XCS40	FPGA	80 MHz	N/A	Not reported	Grayscale images	Invisible	Not reported
Seo and Kim (2003)	VHDL	APEX20KC EP20K400CF672C7 Altera kit	FPGA	82 MHz	N/A	32.3	Grayscale images	Invisible	robust
De Strycker et al. (2000)	C code compiler	TriMedia TM- 1000 (Philips Semiconductors) processor	DSP	100 MHz	N/A	N/A	Images and videos	Invisible	Robust
Cassuto et al. (2000)	MATLAB	TMS 320C5410	DSP	N/A	N/A	21	Audios	Not reported	Robust

ownership. In transform domain watermarking, the transform coefficients of host images are modified by the watermark. The hardware-based implementation of this technique is more complex than the spatial domain technique as it requires additional hardware blocks and resources for the implementation of forward and inverse transforms; where in the parallel, pipelined architectures suit best. There are also approaches in which finding complex operations such as transform/inverse/extraction is done in MATLAB, while other steps of the algorithm are done by FPGA hardware. Many prototypes of digital image watermarking for secure digital cameras are also available, with major components being the image sensor, Analog-to-Digital (A/D) convertor, watermarking unit, temporary memory, flash memory, controller unit, and Liquid Crystal Display (LCD) panel.

4.4 CHALLENGES AND FUTURE DIRECTIONS

Hardware-based implementations are developed for only a few watermarking techniques, such as LSB substitution, SS, additive technique, and DCT/DWT-based techniques using either FPGA boards or DSP kits. Many advanced software watermarking techniques based on channel coding algorithms, phase congruency, singular value decomposition, curvelet transform, contourlet transform, ridgelet transform, hybrid combinations, machine learning, and bio-inspired algorithms are available without hardware-based implementations. Thus, new hardware-based implementation approaches are required for these types of advanced watermarking techniques. There is also a requirement for the development of efficient hardware-based watermarking architectures that operate at a higher frequency and involve fewer resources and computations, while maintaining good imperceptibility and robustness against different types of attacks. The design of prototypes for digital image watermarking using XSG demands for time, area. and cost reduction, and optimization in resource utilization and latency, all with an ease in the compatibility of integrating and configuring the devices.

Some hardware-based watermarking implementations are hardware–software co-designed to save power consumption and time. Some operations which demand high performance are designed in hardware, while the ones which are computationally expensive are implemented in software. These kinds of designs result in a trade-off between hardware and software-based implementations, which can be improved in future.

REFERENCES

AlAli, Mohammad I., Khaldoon M. Mhaidat, and Inad A. Aljarrah. "Implementing image processing algorithms in FPGA hardware." In *IEEE Jordan Conference on Applied Electrical Engineering and Computing Technologies (AEECT'13)*, pp. 1–5. IEEE, 2013.

Cassuto, Yuval, Michael Lustig, and Shay Mizrachy. "Real-time digital watermarking system for audio signals using perceptual masking." In *Prize Wnning Project in the TI (Texas Instruments) DSP and Analog Challenge*, Dallas, 2000.

De Strycker, Lieven, Pascale Termont, Jan Vandewege, J. Haitsma, A. A. C. M. Kalker, Marc Maes, and Geert Depovere. "Implementation of a real-time digital watermarking process for broadcast monitoring on a TriMedia VLIW processor." *IEE Proceedings-Vision, Image and Signal Processing* 147(4): 371–376, 2000.

Do Vinh, Quang, and Insoo Koo. "FPGA Implementation of LSB-based Steganography." *Journal of Information and Communication Convergence Engineering* 15(3): 151–159, 2017.

Garimella, Annajirao, M. V. V. Satyanarayana, P. S. Murugesh, and U. C. Niranjan. "ASIC for digital color image watermarking." In *3rd IEEE Signal Processing Education Workshop, 2004 IEEE 11th, Digital Signal Processing Workshop*, pp. 292–296. IEEE, 2004.

Garimella, Annajirao, M. V. V. Satyanarayana, R. Satish Kumar, P. S. Murugesh, and U. C. Niranjan. "VLSI implementation of online digital watermarking technique with difference encoding for 8-bit gray scale images." In *Proceedings of 16th International Conference on VLSI Design*, pp. 283–288. IEEE, 2003.

Ghosh, Sudip, Nachiketa Das, Subhajit Das, Santi P. Maity, and Hafizur Rahaman. "FPGA and SoC based VLSI architecture of reversible watermarking using rhombus interpolation by difference expansion." In *India Conference (INDICON), 2014 Annual IEEE*, pp. 1–6. IEEE, 2014.

Ghosh, Sudip, Somsubhra Talapatra, Navonil Chatterjee, Santi P. Maity, and Hafizur Rahaman. "FPGA based implementation of embedding and decoding architecture for binary watermark by spread spectrum scheme in spatial domain." *Bonfring International Journal of Advances in Image Processing* 2(4): 1–8, 2012.

Harini, V., and V. Vijayaraghavan, "FPGA implementation of secret data sharing through image by using LWT and LSB steganography technique." *International Journal of Engineering Science* 6(7): 12902–12905, 2017.

Hsiao, Shen-Fu, Yor-Chin Tai, and Kai-Hsiang Chang. "VLSI design of an efficient embedded zerotree wavelet coder with function of digital watermarking." *IEEE Transactions on Consumer Electronics* 46(3): 628–636, 2000.

Joshi, Amit, Vivekanand Mishra, and R. M. Patrikar. "Real time implementation of digital watermarking algorithm for image and video application." In *Watermarking-Volume 2*. InTech, 2012.

Karmani, Sourour, Ridha Djemal, and Rached Tourki. "Efficient hardware architecture of 2D-scan-based wavelet watermarking for image and video." *Computer Standards & Interfaces* 31(4): 801–811, 2009.

Karthigaikumar, P., and K. Baskaran. "FPGA implementation of High Speed Low Area DWT based invisible image watermarking algorithm." *Procedia Engineering* 30: 266–273, 2012.

Khan, Tariq M., D. G. Bailey, Mohammad AU Khan, and Yinan Kong. "Real-time edge detection and range finding using FPGAs." *Optik-International Journal for Light and Electron Optics* 126(17): 1545–1550, 2015.

Kothari, Ashish M., Vedvyas Dwivedi, and Rohit M. Thanki. *Watermarking Techniques for Copyright Protection of Videos*. Springer, Cham, 2018.

Lad, T. C., A. D. Darji, S. N. Merchant, and A. N. Chandorkar. "VLSI implementation of wavelet based robust image watermarking chip." In *2011 International Symposium on Electronic System Design*, pp. 56–61. IEEE, 2011.

Lakshmi, H. R., and B. Surekha. "Asynchronous implementation of reversible image watermarking using mousetrap pipelining." In *IEEE 6th International Conference on Advanced Computing (IACC'16)*, pp. 529–533. IEEE, 2016.

Lakshmi, H. R., B. Surekha, and S. Viswanadha Raju. "Real-time implementation of reversible watermarking." In *Intelligent Techniques in Signal Processing for Multimedia Security*, pp. 113–132. Springer, Cham, 2017.

Langelaar, Gerhard C., Iwan Setyawan, and Reginald L. Lagendijk. "Watermarking digital image and video data. A state-of-the-art overview." *IEEE Signal Processing Magazine* 17(5): 20–46, 2000.

Lim, Hyun, Soon-Young Park, Seong-Jun Kang, and Wan-Hyun Cho. "FPGA implementation of image watermarking algorithm for a digital camera." In *IEEE Pacific Rim Conference on Communications, Computers and signal Processing(PACRIM'03)*, vol. 2, pp. 1000–1003. IEEE, 2003.

Lin, Fei-yu, Wei-ming Qiao, Yan-yu Wang, Tai-lian Liu, Jin Fan, and Jian-chuan Zhang. "Efficient WCDMA digital down converter design using system generator." In *International Conference on Space Science and Communication, 2009. IconSpace*, pp. 89–92. IEEE, 2009.

Madanayake, H. L. P. Arjuna, and Len T. Bruton. "A systolic-array architecture for first-order 3-D IIR frequency-planar filters." *IEEE Transactions on Circuits and Systems I: Regular Papers* 55(6): 1546–1559, 2008.

Maes, Maurice, Ton Kalker, J-PMG Linnartz, Joop Talstra, F. G. Depovere, and Jaap Haitsma. "Digital watermarking for DVD video copy protection." *IEEE Signal Processing Magazine* 17(5): 47–57, 2000.

Maity, Hirak Kumar, and Santi P. Maity. "FPGA Implementation for Modified RCM-RW on Digital Images." *Journal of Circuits, Systems and Computers* 26(03): 1750044, 2017.

Maity, Hirak Kumar, and Santi P. Maity. "FPGA implementation of reversible watermarking in digital images using reversible contrast mapping." *Journal of Systems and Software* 96: 93–104, 2014.

Maity, Santi P., and Malay K. Kundu. "Distortion free image-in-image communication with implementation in FPGA." *AEU-International Journal of Electronics and Communications* 67(5): 438–447, 2013.

Maity, Santi P., Ayan Banerjee, Abu Abhijit, and Malay K. Kundu. "VLSI design of spread spectrum image watermarking." In *13th National Conference on Communications (NCC'07)* Indian Institute of Technology, Kanpur, India, January, pp. 26–28. 2007.

Maity, Santi P., Malay K. Kundu, and Seba Maity. "Dual purpose FWT domain spread spectrum image watermarking in real time." *Computers & Electrical Engineering* 35(2): 415–433, 2009.

Mathai, Nebu John, Ali Sheikholeslami, Deepa Kundur. "VLSI implementation of a real-time video watermark embedder and detector." In *ISCAS'03. Proceedings of the 2003 International Symposium on Circuits and Systems, 2003.*, vol. 2, pp. II–II. IEEE, 2003.

Mohanty, Saraju P., and Sridhara Nayak. "FPGA based implementation of an invisible-robust image watermarking encoder." In *Intelligent Information Technology*, pp. 344–353. Springer, Berlin, Heidelberg, 2004.

Mohanty, Saraju P., Elias Kougianos, and Nagarajan Ranganathan. "VLSI architecture and chip for combined invisible robust and fragile watermarking." *IET Computers & Digital Techniques*1 (5): 600–611, 2007.

Mohanty, Saraju P., N. Ranganathan, and Ravi K. Namballa. "VLSI implementation of invisible digital watermarking algorithms towards the development of a secure JPEG encoder." In *IEEE Workshop on Signal Processing Systems, 2003. SIPS 2003*, pp. 183–188. IEEE, 2003.

Mohanty, Saraju P., Nagarajan Ranganathan, and Ravi K. Namballa. "A VLSI architecture for visible watermarking in a secure still digital camera (S/sup 2/DC) design (Corrected)." *IEEE Transactions on Very Large Scale Integration (VLSI) Systems* 13(8): 1002–1012, 2005.

Morita, Y., E. Ayeh, O. B. Adamo, and P. Guturu. "Hardware/software co-design approach for a dct-based watermarking algorithm." In *52nd IEEE International Midwest Symposium on Circuits and Systems, 2009. MWSCAS'09*. pp. 683–686. IEEE, 2009.

Mulani, Altaf O., and P. B. Mane. "Watermarking and cryptography based image authentication on reconfigurable platform." *Bulletin of Electrical Engineering and Informatics* 6(2): 181–187, 2017.

Nayak, Manas Ranjan, Joyashree Bag, Souvik Sarkar, and Subir Kumar Sarkar. "Hardware implementation of a novel water marking algorithm based on phase congruency and singular value decomposition technique." *AEU-International Journal of Electronics and Communications* 71: 1–8, 2017.

Nelson, Graham R., Graham A. Jullien, and Orly Yadid-Pecht. "CMOS image sensor with watermarking capabilities." In *IEEE International Symposium on Circuits and Systems, 2005. ISCAS 2005*, pp. 5326–5329. IEEE, 2005.

Pemmaraju, Manasa, Sai Chand Mashetty, Srinivas Aruva, Mohanshankar Saduvelly, and Bharat Babu Edara. "Implementation of image fusion based on wavelet domain using FPGA." In *International Conference on Trends in Electronics and Informatics (ICEI'17)*, pp. 500–504. IEEE, 2017.

Petitjean, Guillaume, Jean-Luc Dugelay, Sophie Gabriele, Christian Rey, and Jean Nicolai. "Towards real-time video watermarking for system-on-chip." In *Proceedings of IEEE International Conference on Multimedia and Expo(ICME'02).*, vol. 1, pp. 597–600. IEEE, 2002.

Powar, Prachi V., and S. S. Agrawal. "Design of digital video watermarking scheme using MATLAB Simulink." *International Journal of Research in Engineering and Technology* 2(5): 826–830, 2013.

Que, Zhiqiang, Yongxin Zhu, Xuan Wang, Jibo Yu, Tian Huang, Zhe Zheng, Li Yang, Feng Zhao, and Yuzhuo Fu. "Implementing medical CT algorithms on stand-alone FPGA based systems using an efficient workflow with SysGen and simulink." In *IEEE 10th International Conference on Computer and Information Technology (CIT'10)*, pp. 2391–2396. IEEE, 2010.

Rahimunnisa, K., P. Karthigaikumar, Soumiya Rasheed, J. Jayakumar, and S. Suresh Kumar. "FPGA implementation of AES algorithm for high throughput using folded parallel architecture." *Security and Communication Networks* 7(11): 2225–2236, 2014.

Rajagopalan, Sundararaman, Pakalapati JS Prabhakar, Mucherla Sudheer Kumar, N. V. M. Nikhil, Har Narayan Upadhyay, J. B. B. Rayappan, and Rengarajan Amirtharajan. "MSB based embedding with integrity: An adaptive RGB Stego on FPGA platform." *Inform. Technol. J* 13: 1945–1952, 2014.

Roy, Sonjoy Deb, Xin Li, Yonatan Shoshan, Alexander Fish, and Orly Yadid-Pecht. "Hardware implementation of a digital watermarking system for video authentication." *IEEE Transactions on Circuits and Systems for Video Technology* 23(2): 289–301, 2013.

Samanta, D., A. Basu, T. S. Das, V. H. Mankar, Ankush Ghosh, Manish Das, and Subir K. Sarkar. "SET based logic realization of a robust spatial domain image watermarking." In *International Conference on Electrical and Computer Engineering(ICECE'8)*, pp. 986–993. IEEE, 2008.

Seo, Young-Ho, and Dong-Wook Kim. "Real-time blind watermarking algorithm and its hardware implementation for motion JPEG2000 image codec." In *ESTImedia*, pp. 88–93. 2003.

Shah, D. A. "Authentication of images." Thesis for University Honors Program, Northern Illinois University, DeKalb, Illinois, 2017.

Shet, K. Sathish, A. R. Aswath, M. C. Hanumantharaju, and Xiao-Zhi Gao. "Design and development of new reconfigurable architectures for LSB/multi-bit image steganography system." *Multimedia Tools and Applications* 76(11): 13197–13219, 2017.

Shih, Frank Y. *Digital Watermarking and Steganography: Fundamentals and Techniques*. CRC Press, Boca Raton, FL, 2017.

Tsai, S. E., and S. M. Yang. "A fast DCT algorithm for watermarking in digital signal processor." *Mathematical Problems in Engineering* 2017, 2017.

Tsai, T. H., and C. Y. Lu, "A systems level design for embedded watermark technique using DSC systems." In *Proceedings of the IEEE International Workshop on Intelligent Signal Processing and Communication Systems*, November, pp. 20–23, 2001.

Tsai, Tsung-Han, and Chih-Yen Wu. "An implementation of configurable digital watermarking system in MPEG video encoder." In *IEEE International Conference on Consumer Electronics(ICCE'03)*, pp. 216–217. IEEE, 2003.

Venugopala, P. S., H. Sarojadevi, and Niranjan N. Chiplunkar. "An approach to embed image in video as watermark using a mobile device." *Sustainable Computing: Informatics and Systems* 15: 82–87, 2017.

Zavaleta, Z. Jezabel Guzmán, Claudia Feregrino Uribe, and René Cumplido. "A reversible data hiding algorithm for radiological medical images and its hardware implementation." In *International Conference on Reconfigurable Computing and FPGAs*, pp. 444–449. IEEE, 2008.

CHAPTER 5

Applied Examples and Future Prospectives

ANALOG WATERMARKS ARE MOSTLY visible and found on currency, postal stamps, passports, bank notes, paper documents, etc. In the early days of invention, digital watermarking use was limited to proving the copyright owner of intellectual property. Later, with the rapid use of the Internet, digital watermarking found its employability in a wide range of applications, such as e-advertising, e-delivery, e-governance, e-education, e-voting, military, covert communication, telemedicine, web publishing, digital forensics, digital library, robotics, hardware/chip protection, device control, legacy enhancement, media file archiving, computer programs, real-time audio/video, etc.

Digital watermarking protects information and is capable of detecting illegal use of proprietary data. A robust invisible digital watermark is imperceptible information hidden intentionally by the owner of the image into the host image so that the extraction of information at a later stage helps for various purposes, such as copyright protection, authentication, ownership proof, copy and usage control, fingerprinting, tamper detection/evidence, transaction tracking, and broadcast monitoring (Surekha and Swamy, 2012; Surekha et al., 2012, 2016). The requirements of any watermarking system depend on the type of application. This chapter discusses various applications and possible requirements of watermarking in the areas of telemedicine, remote sensing/military, and industry.

5.1 APPLICATIONS OF WATERMARKING

A DRM system recommended watermarking for various applications with variable levels of robustness and other requirements. A few applications of digital watermarking are discussed in this section.

- **Owner identification**

  Proof of ownership is achieved by embedding watermarks in the images published on the Web, which probably conveys information about the author and/or source of the image to convey the ownership and to claim/resolve the ownership rights. This application requires the watermark to be robust, unambiguous, and resistant to the addition of additional watermarks by adversaries. Ex: Digimarc System.

- **Copy protection**

  Illegal copying/downloading of images over networks can be prevented by the application of copy never/copy once watermarking. Ex: DVD system.

- **Fingerprinting/Copy or usage control**

  Copy protection is extremely difficult in open and distributive systems. Illegal violation of licence agreements/copying/usage/redistribution of images by the authorized users over open networks can be prevented or identified by fingerprinting each image copy with a unique identifier. The main requirement of fingerprinting is robustness to collusion attacks. Ex: Divx System, theater identification from pirated copy.

- **Data authentication and integrity/Tamper detection**

  Data tampering is very easy and should be detected particularly for digital forensic applications. The signatures/header/metadata does not indicate the changes done to the images. Fragile watermarking, where the watermark disappears when data is modified, exhibits very low robustness and is preferred in identifying if the data is intact. Tamper detection is crucial when dealing with applications involving medical, military, and satellite imagery.

- **Broadcast monitoring**

  Watermarking can be used to monitor the broadcasting of images as a solution to royalty and marketing issues.

- **Transaction tracking**

  The invisible watermarks can be inserted into images to keep track of transactions in e-commerce applications. It is a type of remote triggering during distribution.

- **Content archiving/Filtering/Classification/E-commerce**

  A unique digital watermark can be inserted into images so as to get them identified easily for classification and archiving purpose.

- **ID security**

  Identity cards and documents are often stolen, copied, counterfeited, altered, or regenerated by attackers. Identity cards can be scanned to detect watermarks for authentication, detection of fraud, forensic analysis, and tracking.

5.2 WATERMARKING IN TELEMEDICINE

Rapid growth in Internet and communication technologies and the deployment of digital picture archiving and communication systems (PACS), hospital information systems (HIS), and radiology information systems (RIS), allowed the invention of telemedicine, which includes sharing and exchange of expertise, radiological images, and electronic patient records (EPR), among several entities (Thanki and Borra, 2018).

EPRs and medical media are digitally generated and transmitted over open networks as part of telediagnosis, teleconsulting, telesurgery, home monitoring, emergency treatment, and medical education. These documents are very personal and include crucial information about patients, hospitals, diagnostic centers, doctors, etc., which are to be digitally authenticated and verified for assuring confidentiality, authentication, and data integrity. With present-day technological advancements, it is very easy to intercept, copy, track, and modify the medical data. Hence, the secure transfer of data plays a crucial role in facilitating the benefits of the telemedicine (Thanki and Borra, 2018; Dey et al., 2016; Bose et al., 2014; Chowdhury et al., 2014).

The digital imaging and communications in medicine (DICOM) standard recommended cryptographic techniques, such as encryption, digital signatures, and hashing functions, for providing integrity and authenticity (Thanki et al., 2018). Since cryptography cannot provide protection after decryption, data hiding and watermarking methods are suggested, particularly for identity authentication, privacy concerns, and ownership verification. The digital signature stored in the header of the DICOM image can be easily removed or modified by adversaries, thereby affecting its authenticity and integrity. This is due to the fact that DICOM has limited its recommendations to the confidentiality of the header but does not recommend any for ensuring the authenticity or integrity of the header. Apart from robust watermarks, fragile watermarks can be used for tamper detection and data integrity. Hybrid methods, such as crypto-watermarks, address most of the security issues and ensure that the exchanged medical images belong to the right person and organization, that the data is not manipulated, and that the data is accessed by authorized people.

Most of the medical image watermarking algorithms defined in the literature are irreversible, lossy in nature, and permanently degrade the crucial content useful for diagnosis while embedding the watermarks (Borra et al., 2018; Chatterjee et al., 2018; Dey et al., 2012, 2017a; Banerjee et al., 2015; Biswas et al., 2013; Pal et al., 2013; Das et al., 2012). To overcome these limitations, region of non-interest (RONI)–based watermarking techniques and reversible watermarking techniques, which can restore the lost data, are developed (Dey et al., 2017b). Hash functions calculated from the region of interest (ROI) of medical images can also be embedded as an additional watermark. Embedding multiple watermarks in medical images enhances security and flexibility, allows detection of tampering, and reduces the bandwidth and storage requirements. However, embedding of multiple watermarks in a single medical image degrades the quality and affects the image's ability to survive against attacks.

5.3 ROLE OF WATERMARKING IN REMOTE-SENSING MILITARY

Modern remote sensing and space technologies made it possible to acquire lots and lots of images per second. The Earth observation satellites collect data from Earth remotely and transmit the data to an Earth station, which is then processed and maintained in a database for archiving, viewing, and purchasing. These images may further be analyzed and interpreted for decision making in various scientific or commercial applications, such as weather monitoring, area observations, digital elevation models, land cover, vegetation, soil moisturizer, defense and so on.

Satellite images are mostly acquired in multiple spectrums by specially designed sensors, which are very expensive and which involve complex installation procedures and post-processing. Multispectral images are part of big data. Satellite images and aerial imagery are rich sources of information and may even contain confidential data related to military and war fields, apart from commercial data. Information hiding in remote sensing should be handled extremely carefully, as the data is very sensitive. These highly sensitive images are prone to leakage and are to be protected with greater levels of confidentiality and access. It takes more time to encrypt and decrypt these images and the key management is too tedious. Hence, it is very important to protect acquired images from misuse, illegal access, and usage, as well as protecting the copyrights. Secure storage, maintenance, archiving, and transmission of these remotely sensed images from unauthorized personnel is a challenging task. Traditional watermarking techniques developed for images acquired in the visible band cannot be directly applied to satellite images as they are multispectral and hyperspectral images and can even have a raster data structure. Hence, a crypto watermarking technique is most suitable for these images.

The crucial requirements of watermarking techniques when applied to satellite images are imperceptibility, lossless embedding, the capacity to extract the watermark blindly, the ability to watermark region of interest and integrity of data. Robustness to compression, noise, and filtering is sufficient in most of the applications. It is important that the watermark embedding should not affect the image's classification, accuracy, matching, and measurements. The watermark should be able to be detected at multiple resolutions. The reversible watermarking schemes are widely used in military applications where copyright protection or ownership authentication is required. The watermark and the original image are extracted from the watermarked image.

5.4 INDUSTRIAL AND MISCELLANEOUS APPLICATIONS

Product designs and digital media are forms of intellectual property (IP) and are a result of valuable effort. Hence, mechanisms that protect the rights of IP producers and owners are in considerable demand and are of great interest. Physical layout designs that involve routing and placement, are very sensitive, and must be resistant to tampering, as any changes made to them lead to adverse effects on design productivity and on the market. Watermarking addresses IP protection by tracing unauthorized use, duplication, and tracking.

Very often people scan their personal documents and identity cards and make digital copies before saving and carrying. Identity thieves constantly try to misuse personal data. Preparatory activities to thwart such attacks include encryption, digital signatures, access controls, and watermarking by companies issuing the identity cards.

It is necessary that watermarking techniques applied to IP and identity card protection include features such as, invisibility, persistence, data integrity, robustness, enforceability, and security. Furthermore, the watermarking process should retain functional correctness and component protection with a minimal overhead cost.

5.5 FUTURE PROSPECTIVES

A lot of research has been done in the area of digital image watermarking and its application areas in the last twenty years, a major focus being copyright protection and ownership authentication of proprietary images, digital photographs, digital artworks, industrial designs, product models, medical images, remotely sensed images, etc. The development of intelligent watermarking, is adaptable and involves less manual intensiveness. However, some problems still persist and will need to be addressed in the future. These issues are listed below:

- In traditional watermarking methods, which are irreversible, the main challenge is to avoid compromising robustness, capacity, and imperceptibility.

- Watermarking methods that can simultaneously ensure integrity, confidentiality, and authenticity are of great interest.
- The techniques that involve segmentation of an image into region of non-interest (RONI) and region of interest (ROI) have limitations with respect to the choice of region for watermark embedding. If watermarks are embedded in ROI, the crucial information gets degraded. On the other hand, the watermark size should be very small if they are to be embedded in RONI.
- Although reversible watermarking maintains the original quality of images, it has limitations related to payload capacity, computational complexity, and time.
- Standard encryption algorithms cannot be utilized for multispectral images due to their sensitiveness and characteristics. Due to the high-volume of demand, watermarking satellite images is complex in terms of time and key management.
- Watermarking physical designs which are composed of complex structures, such as routing, placement, vertex ordering, and graph coloring is a very challenging task as even a slight change can lead to design failure.
- High-speed real-time watermarking with less computational complexity would be both timely and useful.
- The development of standard databases for testing the performance of image watermarking is required.
- Standardization of digital watermarking is required.
- The development of intelligent watermarking techniques is needed.
- The applicability of deep learning to watermarking needs to be explored.

REFERENCES

Banerjee, Shubhendu, Sayan Chakraborty, Nilanjan Dey, Arijit Kumar Pal, and Ruben Ray. "High payload watermarking using residue number system." *International Journal of Image, Graphics and Signal Processing* 3: 1–8, 2015.

Biswas, Debalina, Poulami Das, Prosenjit Maji, Nilanjan Dey, and Sheli Sinha Chaudhuri. "Visible watermarking within the region of non-interest of medical images based on fuzzy C-means and Harris corner detection." In *Computer Science & Information Technology* 24: 161–168, 2013.

Borra, Surekha, Rohit Thanki, Nilanjan Dey, Komal Borisagar. "Secure transmission and integrity verification of color radiological images using fast discrete curvelet transform and compressive sensing." *Smart Health* 2018.

Bose, Soumyo, Shatadru Roy Chowdhury, Chandreyee Sen, Sayan Chakraborty, Taiar Redha, and Nilanjan Dey. "Multi-thread video watermarking: A biomedical application." In *International Conference on Circuits, Communication, Control and Computing (I4C)*, IEEE. pp. 242–246, 2014.

Chatterjee, Sankhadeep, Rhitaban Nag, Nilanjan Dey, and Amira S. Ashour. "Efficient economic profit maximization: Genetic algorithm based approach." In *Smart Trends in Systems, Security and Sustainability*, pp. 307–318. Springer, Singapore, 2018.

Chowdhury, Shatadru Roy, Ruben Ray, Nilanjan Dey, Sayan Chakraborty, Wahiba Ben Abdessalem Karaa, and Siddhartha Nath. "Effect of demon's registration on biomedical content watermarking." In *International Conference on Control, Instrumentation, Communication and Computational Technologies (ICCICCT'14)* pp. 509–514. IEEE, 2014.

Das, Poulami, Riya Munshi, and Nilanjan Dey. "Alattar's method based reversible watermarking technique of EPR within heart sound in wireless telemonitoring." *Intellectual Property Rights and Patent Laws, IPRPL-2012, GKCEM* 8: 25, 2012.

Dey, Nilanjan, Amira S. Ashour, Sayan Chakraborty, Sukanya Banerjee, Evgeniya Gospodinova, Mitko Gospodinov, and Aboul Ella Hassanien. "Watermarking in biomedical signal processing." In *Intelligent Techniques in Signal Processing for Multimedia Security*, pp. 345–369. Springer, Cham, 2017a.

Dey, Nilanjan, Amira S. Ashour, Fuqian Shi, Simon James Fong, and R. Simon Sherratt. "Developing residential wireless sensor networks for ECG healthcare monitoring." *IEEE Transactions on Consumer Electronics* 63(4): 442–449, 2017b.

Dey, Nilanjan, Shouvik Biswas, Poulami Das, Achintya Das, and Sheli Sinha Chaudhuri. "Lifting wavelet transformation based blind watermarking technique of photoplethysmographic signals in wireless telecardiology." In *World Congress on Information and Communication Technologies (WICT'12)*, pp. 230–235. IEEE, 2012.

Dey, Nilanjan, Soumyo Bose, Achintya Das, Sheli Sinha Chaudhuri, Luca Saba, Shoaib Shafique, Andrew Nicolaides, and Jasjit S. Suri. "Effect of watermarking on diagnostic preservation of atherosclerotic ultrasound video in stroke telemedicine." *Journal of Medical Systems* 40(4): 91, 2016.

Pal, Arijit Kumar, Nilanjan Dey, Sourav Samanta, Achintya Das, and Sheli Sinha Chaudhuri. "A hybrid reversible watermarking technique for color biomedical images." In *IEEE International Conference on Computational Intelligence and Computing Research (ICCIC'13)*, pp. 1–6. IEEE, 2013.

Surekha, B., and G. N. Swamy. "Digital image ownership verification based on spatial correlation of colors." In *IET Conference on Image Processing (IPR 2012)*, London, pp. 1–10, 2012.

Surekha, B., G. Swamy, and K. Rama Linga Reddy. "A novel copyright protection scheme based on visual secret sharing." In *Third International Conference on Computing Communication & Networking Technologies (ICCCNT'12)*, pp. 1–5. IEEE, 2012.

Thanki, Rohit M., Surekha Borra, and Komal R. Borisagar. "A hybrid watermarking technique for copyright protection of medical signals in teleradiology." In *Handbook of Research on Information Security in Biomedical Signal Processing*, pp. 320–349. IGI Global, 2018.

Thanki, Rohit, and Surekha Borra. *Medical imaging and its security in telemedicine applications*. Springer, Berlin, 2018.

CHAPTER 6

Case Study

Watermarking techniques are mainly used for copyright protection or the authentication of multimedia data, such as images, video, and audio (Borra et al., 2018, 2017; Lakshmi and Surekha, 2016; Surekha and Swamy, 2012, 2014, 2016) However, the parameters chosen for optimization depend on the methodology involved in watermark embedding, extraction, and its application. The major requirements of watermarking are imperceptibility, robustness, and security against different kinds of attacks, in which the selection of the gain factor plays a crucial role. Many researchers have used gain factors by a trial-and-error method to meet their transparency requirements of a watermarked image. The manual selection of the gain factor is a time-consuming process and is not a standard approach. The value of the gain factor is also kept constant in many existing schemes in the literature and is not adaptable to various kinds of images and applications (Borra et al., 2018; Thanki et al., 2017). Thus, to make the watermarking scheme application dependent and to support and adjust the gain factors automatically for different host images, some adaptive schemes are required. In this case study, therefore, a popular optimization algorithm, particle swarm optimization (PSO), is combined with a block redundant wavelet transform (RDWT)-based watermarking scheme to obtain optimized gain factors. The flow chart and working of the PSO algorithm is given in Chapter 3. Optimization algorithms, which were discussed in Chapter 3, are used in finding optimized gain factors to give the best results for a particular application. Out of all of these algorithms, PSO algorithm is broadly used for optimization of gain factors in digital image watermarking due to its ease of understanding, ease of implementation, and less computational time compared to other algorithms.

In this chapter an optimized and blind invisible robust watermarking scheme is developed based on block redundant wavelet transform (RDWT) and PSO (Thanki et al., 2017) to overcome some limitations of the existing techniques. The RDWT domain coefficients are modified for watermark insertion based on watermark bits using noise sequences. The usage of the RDWT provides better imperceptibility and payload capacity to watermarking compared to existing simple wavelet transform-based schemes. Employing PSO improves imperceptibility and robustness. The experimental results and comparative analysis with an existing technique in the literature are given to show that the presented bio-inspired watermarking technique performs well in terms of imperceptibility and payload capacity.

6.1 EMBEDDING ALGORITHM

The proposed scheme is based on block RDWT and PSO, in which the monochrome watermark image is inserted directly into detail wavelet subbands like LH, HL, and HH of the host image. Each wavelet subband is divided into nonoverlapping blocks. For the embedding of watermark, two uncorrected noise sequences for watermark bits "0" and bit "1" are generated. Each noise sequence modifies the coefficients of the corresponding subband of the host image with the help of the optimized gain factor obtained using the PSO algorithm. The simplified block diagram of the proposed digital image watermarking scheme is shown in Figure 6.1. The steps involved in the watermark embedding process are given below:

1. Decompose the host image C into RDWT wavelet subbands: LL, LH, HL, and HH.
2. Convert the monochrome image w into a binary sequence.
3. Divide the detail wavelet subbands, LH, HL, and HH, into nonoverlapped blocks.
4. Generate two uncorrelated noise sequences using a noise generator, each of an equal size to the nonoverlapped block.

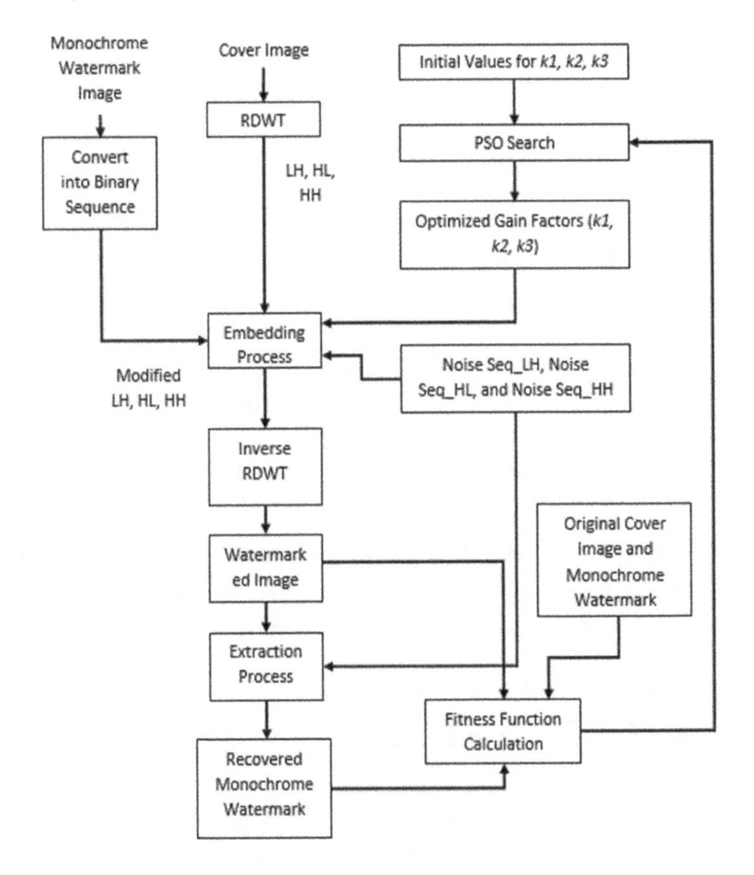

FIGURE 6.1 Block diagram of proposed optimized watermarking scheme.

5. Modify the coefficients of detail wavelet subbands of the host image based on watermark bits and gain factors obtained from the PSO search algorithm using Equations.6.1 and 6.2:
   - If watermark bit == 0 then

$$\begin{aligned} &\text{Modified\_LH} = \text{LH} + k1 * \text{Noise\_Seq0} \\ &\text{Modified\_HL} = \text{HL} + k2 * \text{Noise\_Seq0} \\ &\text{Modified\_HH} = \text{LH} + k3 * \text{Noise\_Seq0} \end{aligned} \tag{6.1}$$

   - else if watermark bit == 1 then

$$\begin{aligned} &\text{Modified\_LH} = \text{LH} + k1 * \text{Noise\_Seq1} \\ &\text{Modified\_HL} = \text{HL} + k2 * \text{Noise\_Seq1} \\ &\text{Modified\_HH} = \text{LH} + k3 * \text{Noise\_Seq1} \end{aligned} \tag{6.2}$$

   Where Modified_LH, Modified_HL, and Modified_HH represent the modified coefficients of wavelet subbands; LH, HL, and HH represents original coefficients of wavelet subbands; $k1$, $k2$, and $k3$ are optimized gain factors; and Noise_Seq0 and Noise_Seq1 are noise sequences corresponding to watermark bit 0 and 1.

   - Repeat the procedure for all blocks of the host image, to hide all watermark bits.

6. Apply single-level inverse RDWT on the modified detail wavelet subbands along with original approximation subband to get the watermarked image WA.

6.2 EXTRACTION ALGORITHM

The process of watermark extraction is given below:

1. Decompose the watermarked image/test image WA into RDWT wavelet subbands: approximation subband LL and detail subbands, such as LH, HL, and HH.

2. Divide the detail wavelet subbands, LH, HL, and HH, into nonoverlapped blocks.
3. Consider uncorrelated noise sequences, which are generated during the embedding algorithm, which are maintained in secret by the owner of the image.
4. Recover the watermark bit from the detail wavelet coefficients based on the following conditions defined by Equations 6.3 and 6.4:

$$\begin{aligned}
&\text{Seq}A\text{=corr2(Modified_LH,Noise_Seq0)} \\
&\text{Seq}B\text{=corr2(Modified_HL,Noise_Seq0)} \\
&\text{Seq}C\text{=corr2(Modified_HH,Noise_Seq0)} \\
&\text{Seq0} = (\text{Seq}A + \text{Seq}B + \text{Seq}C)/3
\end{aligned} \tag{6.3}$$

$$\begin{aligned}
&\text{Seq}X = \text{corr2(Modified_LH,Noise_Seq1)} \\
&\text{Seq}Y = \text{corr2(Modified_HL,Noise_Seq1)} \\
&\text{Seq}Z = \text{corr2(Modified_HH,Noise_Seq1)} \\
&\text{Seq1} = (\text{Seq}X + \text{Seq}Y + \text{Seq}Z)/3
\end{aligned} \tag{6.4}$$

5. If $\text{Seq0} > \text{Seq1}$ then set watermark bit as 0, else as bit 1.
6. Reshape the extracted sequence into a matrix form to detect the watermark w'.

6.3 SIMULATION RESULTS

The performance of any invisible watermarking scheme mostly depends on the value of the gain factor. Larger gain factors degrade the imperceptibility of the watermarked image but improve the quality of the recovered watermark and vice-versa. The fitness function used in this algorithm is a function of PSNR and NC values and is given by Equation (6.5):

$$\text{fitness} = \text{PSNR}(C, \text{WA}) + (100 * \text{NC}(w, w')) \tag{6.5}$$

where PSNR and NC represents peak signal-to-noise ratio and normalized correlation. Variables C and WA indicate host and watermarked images, and w and w' are original and extracted watermark images. The optimized gain factors achieved ($k1$, $k2$, $k3$) using PSO algorithm for the proposed watermarking scheme are given in Table 6.1. The experiments and comparative analysis of the proposed scheme are done considering three gray host images, Cameraman, Lena and Goldhill, each of size 512×512 pixels (Figure 6.2 a–c), considering a 64×64 pixels monochrome logo as a watermark (Figure 6.2 d). The RDWT subbands are divided into 4096 nonoverlapping blocks each of size 8×8. The maximum number of watermark bits inserted into a host image is calculated using the equation:

$$\text{Maximum\_Watermark} = \frac{M \times N}{\text{Blocksize}^2} \tag{6.6}$$

Where M is row size and N is column size of a host image, respectively.

The proposed scheme can embed a maximum watermark of size 64×64 pixels. The maximum payload of the proposed scheme for the given set of host images is then 1 bit for every 64 pixels of the host image. The results of applying watermark embedding and extraction using proposed optimized gain factors are shown in Figure 6.3.

TABLE 6.1 Optimized Gain Factor Values Using PSO Algorithm in Proposed Scheme

Range of Gain Factors	k_1	k_2	k_3
0.0–1.0	0.5483	0.6346	0.8362
0.0–2.0	1.5718	1.8092	1.6667
0.0–3.0	2.0803	2.7933	2.6394
0.0–4.0	3.8015	3.8167	3.8451
0.0–8.0	7.6857	6.2683	5.6797
0.0–10.0	8.0542	9.8810	7.6335
0.0–50.0	42.8845	49.0747	44.0159
0.0–150.0	143.4951	148.4743	142.2608
0.0–250.0	233.1310	225.7728	218.1750

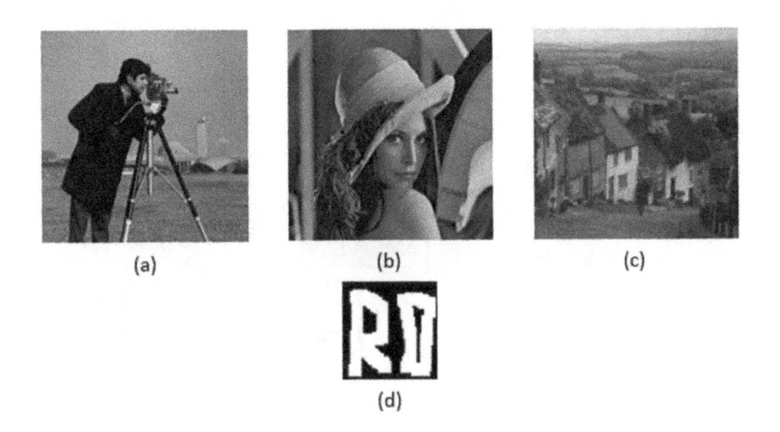

FIGURE 6.2 (a)–(c) Test host images and (d) watermark.

The measured performance metrics of the proposed scheme for different optimized gain factors and grayscale host images are given in Table 6.2. The comparative analysis of imperceptibility achieved with the proposed and existing schemes for different grayscale host images are given in Table 6.3. The numerical values indicate better imperceptibility of proposed scheme compared to many existing schemes. The results also show that as robustness increases with gain factor, imperceptibility decreases. The subjective results of watermark extraction after application of different watermarking attacks on the watermarked images are shown in Figure 6.4. The corresponding NC values are tabulated in Table 6.4. These results indicate that the proposed scheme is robust to most of the attacks, except motion blurring, rotation, and scaling. The figures are generated using gain factors as follows: $k1 = 233.1310$, $k2 = 225.7728$, and $k3 = 218.1750$.

6.4 MAIN FEATURES OF PROPOSED SCHEME

The main features of the proposed optimized watermarking scheme include:

- Reduction in the trade-off between imperceptibility and robustness, existing in the traditional watermarking approaches and added optimization mechanism.

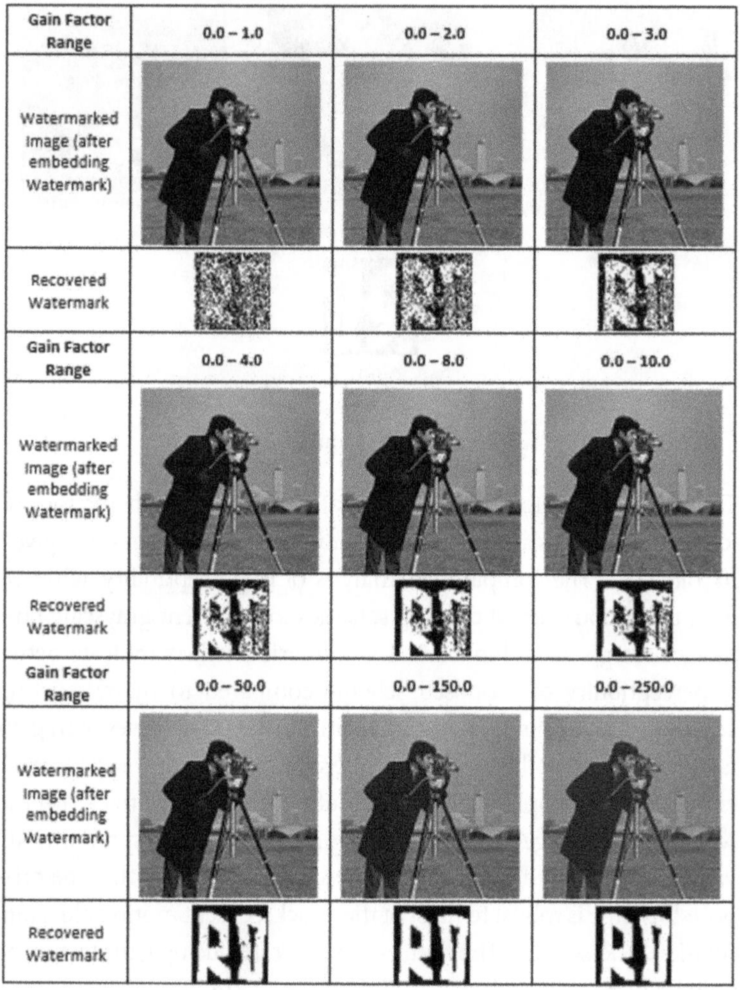

Gain Factor Range	0.0 – 1.0	0.0 – 2.0	0.0 – 3.0
Watermarked Image (after embedding Watermark)			
Recovered Watermark			
Gain Factor Range	0.0 – 4.0	0.0 – 8.0	0.0 – 10.0
Watermarked Image (after embedding Watermark)			
Recovered Watermark			
Gain Factor Range	0.0 – 50.0	0.0 – 150.0	0.0 – 250.0
Watermarked Image (after embedding Watermark)			
Recovered Watermark			

FIGURE 6.3 Watermarked images and recovered watermark image using proposed scheme for grayscale cameraman image.

TABLE 6.2 Quality Measures of Proposed Scheme for Chosen Host Images

Range of Gain Factors	PSNR (dB)	NC	Fitness	T_{EMB} (s)	T_{EXT} (s)
Cameraman Host Image					
0.0–1.0	65.02	0.6105	126.07	1.4655	1.6329
0.0–2.0	57.14	0.6947	126.61	1.4925	1.5458
0.0–3.0	53.66	0.7707	130.73	1.5674	1.5924
0.0–4.0	50.03	0.8435	134.38	1.3832	1.7518
0.0–8.0	45.27	0.9062	135.89	1.5268	1.5565
0.0–10.0	42.99	0.9423	137.22	1.3550	1.7260
0.0–50.0	30.32	0.9922	129.54	1.0581	1.5905
0.0–150.0	20.24	0.9995	120.19	1.7243	1.6175
0.0–250.0	16.38	1.0000	116.38	1.7458	1.5355
Lena Host Image					
0.0–1.0	66.80	0.6238	129.18	1.5732	1.5369
0.0–2.0	58.92	0.7140	130.32	1.5834	1.5662
0.0–3.0	55.43	0.7808	133.51	1.4037	1.5990
0.0–4.0	51.81	0.8403	135.84	1.4061	1.7324
0.0–8.0	47.05	0.8824	135.29	1.4973	1.5379
0.0–10.0	44.78	0.9126	136.04	1.5061	1.5904
0.0–50.0	28.53	0.9991	128.44	1.7832	1.5359
0.0–150.0	18.46	1.0000	118.46	1.7414	1.8264
0.0–250.0	14.60	1.0000	114.60	1.7394	1.5727
Goldhill Host Image					
0.0–1.0	67.31	0.5314	120.45	1.4302	1.5276
0.0–2.0	63.91	0.5327	117.18	1.4845	1.5934
0.0–3.0	55.94	0.5735	113.29	1.3739	1.9731
0.0–4.0	52.32	0.6005	112.37	1.3436	1.4982
0.0–8.0	47.56	0.6490	112.46	1.4583	1.5496
0.0–10.0	45.29	0.7103	116.32	1.4281	1.6299
0.0–50.0	30.82	0.9744	128.26	1.7482	1.6907
0.0–150.0	20.75	0.9986	120.61	1.7539	1.5315
0.0–250.0	16.89	0.9995	116.84	1.8148	1.7301

TABLE 6.3 PSNR Comparison of the Proposed and Existing Schemes

Scheme	Size of Host Image	Size of Watermark	Maximum PSNR (dB)
Li and Wang (2007)	256×256	128×72	38.02
Rohani and Avanaki (2009)	256×256	40×40	37.48
Aslantas et al. (2009)	256×256	64×64	55.76
Findik et al. (2010)	512×512	64×64	42.76
Fakhari et al. (2011)	512×512	256×256	51.55
Wang et al. (2011)	512×512	Not reported	47.40
Bedi et al. (2012)	512×512	90×90	39.42
Run et al. (2012)	512×512	256×256	33.93
Golshan and Mohammadi (2013)	256×256	64×64	45.62
Bedi et al. (2013)	512×512	128×128	46.67
Naheed et al. (2014)	Not reported	Not reported	40.00
Peng et al. (2014)	512×512	Not reported	41.74
Li (2014)	512×512	32×64	42.48
Thakkar and Srivastava (2017)	512×512	64×64	50.68
Sanku et al. (2018)	512×512	64×64	40.91
Proposed scheme	512×512	64×64	67.31

- Utilization of properties of RDWT and PSO to overcome some limitations of standard watermarking procedure such as manual selection of gain factor.
- Decomposition of images of size $M \times N$ into its wavelet sub-bands each of size $M \times N$. This allows embedding of large watermark in contrast to Thakkar scheme (Thakkar and Srivastava, 2017). The RDWT also eliminates the sampling process and overcomes limitations of DWT, including time shift variance.
- Blind extraction of a watermark image (Fakhari et al., 2011; Run et al., 2012; Golshan and Mohammadi, 2013; Verma et al., 2016; Sanku et al., 2018).
- The scheme is robust in nature and performs better than existing Thakkar scheme (Thakkar and Srivastava, 2017) in terms of imperceptibility and payload capacity.

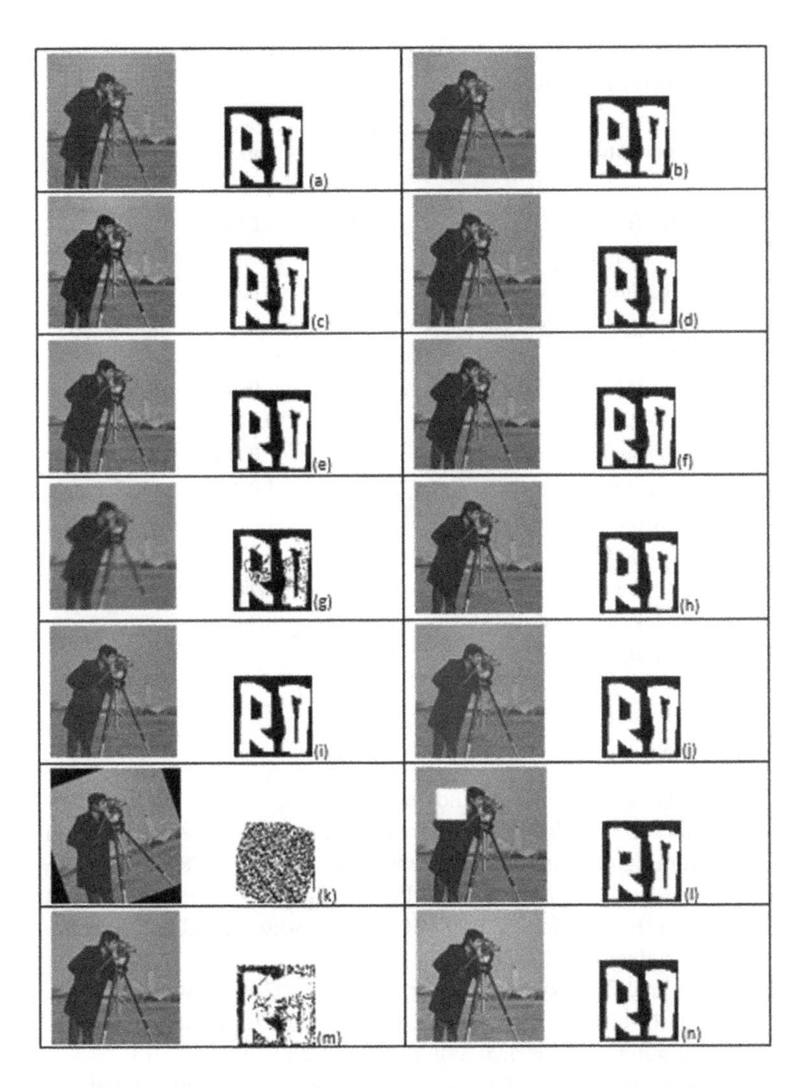

FIGURE 6.4 Watermarked images and recovered watermark image using proposed scheme under different watermarking attacks for gray-scale cameraman image: (a) JPEG (Q = 80), (b) JPEG (Q = 25), (c) Median Filtering (3×3), (d) Gaussian Noise (= 0.005), (e) Salt & Pepper Noise (= 0.005), (f) Speckle Noise (= 0.005), (g) Motion Blurring, (h) Gaussian Blurring, (i) Sharping, (j) Histogram Equalization, (k) Rotation (20), (l) Cropping, (m) Scaling (512-256-512), (n) Intensity Adjustment.

TABLE 6.4 NC Values of Proposed Scheme under Different Watermarking Attacks

Type of Attack	Cameraman Image	Lena Image	Goldhill Image
JPEG (Q = 80)	1.0000	1.0000	0.9995
JPEG (Q = 25)	1.0000	1.0000	0.9995
Median filtering (3 × 3)	0.9982	1.0000	0.9973
Gaussian noise (σ = 0.005)	1.0000	1.0000	0.9995
Salt & pepper noise (σ = 0.005)	1.0000	1.0000	0.9995
Speckle noise (σ = 0.005)	1.0000	1.0000	0.9995
Motion blurring	0.8092	0.8229	0.7643
Gaussian blurring	1.0000	1.0000	0.9995
Sharping	1.0000	1.0000	0.9995
Histogram equalization	1.0000	1.0000	0.9995
Rotation (20°)	0.4998	0.5039	0.5144
Cropping (20%)	0.9995	0.9995	0.9991
Scaling (512 – 256 – 512)	0.9519	0.9620	0.9588
Intensity adjustment	1.0000	1.0000	0.9995

REFERENCES

Aslantas, Veysel, Saban Ozer, and Serkan Ozturk. "Improving the performance of DCT-based fragile watermarking using intelligent optimization algorithms." *Optics Communications* 282(14): 2806–2817, 2009.

Bedi, Punam, Roli Bansal, and Priti Sehgal. "Multimodal biometric authentication using PSO based watermarking." *Procedia Technology* 4: 612–618, 2012.

Bedi, Punam, Roli Bansal, and Priti Sehgal. "Using PSO in a spatial domain-based image hiding scheme with distortion tolerance." *Computers & Electrical Engineering* 39(2): 640–654, 2013.

Borra, Surekha, Rohit Thanki, Nilanjan Dey, and Komal Borisagar. "Secure transmission and integrity verification of color radiological images using fast discrete curvelet transform and compressive sensing." *Smart Health*, 2018.

Borra, Surekha, H. Lakshmi, N. Dey, A. Ashour, and F. Shi. "Digital image watermarking tools: State-of-the-art." In *Information Technology and Intelligent Transportation Systems: Proceedings of the 2nd International Conference on Information Technology and Intelligent Transportation Systems*, Xi'an, China. vol. 296, p. 450. 2017.

Fakhari, Pegah, Ehsan Vahedi, and Caro Lucas. "Protecting patient privacy from unauthorized release of medical images using a bio-inspired wavelet-based watermarking approach." *Digital Signal Processing* 21(3): 433–446, 2011.

Findik, Oğuz, İsmail Babaoğlu, and Erkan Ülker. "A color image watermarking scheme based on hybrid classification method: particle swarm optimization and k-nearest neighbor algorithm." *Optics Communications* 283(24): 4916–4922, 2010.

Golshan, F., and K. Mohammadi. "A hybrid intelligent SVD-based perceptual shaping of a digital image watermark in DCT and DWT domain." *The Imaging Science Journal* 61(1): 35–46, 2013.

Lakshmi, H. R., and B. Surekha. "Asynchronous implementation of reversible image watermarking using mousetrap pipelining." In *IEEE 6th International Conference on Advanced Computing (IACC'16)*, pp. 529–533. IEEE, Bhimavaram, India, 2016

Li, Jianzhong. "An optimized watermarking scheme using an encrypted gyrator transform computer generated hologram based on particle swarm optimization." *Optics Express* 22(8): 10002–10016, 2014.

Li, Xiaoxia, and Jianjun Wang. "A steganographic method based upon JPEG and particle swarm optimization algorithm." *Information Sciences* 177(15): 3099–3109, 2007.

Naheed, Talat, Imran Usman, Tariq M. Khan, Amir H. Dar, and Muhammad Farhan Shafique. "Intelligent reversible watermarking technique in medical images using GA and PSO." *Optik-International Journal for Light and Electron Optics* 125(11): 2515–2525, 2014.

Peng, Hong, Jun Wang, Mario J. Pérez-Jiménez, and Agustín Riscos-Núñez. "The framework of P systems applied to solve optimal watermarking problem." *Signal Processing* 101: 256–265, 2014.

Rohani, Mohsen, and Alireza Nasiri Avanaki. "A watermarking method based on optimizing SSIM index by using PSO in DCT domain." In *14th International CSI Computer Conference*, Tehran, Iran, 2009.

Run, Ray-Shine, Shi-Jinn Horng, Jui-Lin Lai, Tzong-Wang Kao, and Rong-Jian Chen. "An improved SVD-based watermarking technique for copyright protection." *Expert Systems with Applications* 39(1): 673–689, 2012.

Sanku, Deepika, Sampath Kiran, Tamirat Tagesse Takore, and P. Rajesh Kumar. "Digital image watermarking in RGB host using DWT, SVD, and PSO techniques." In *Proceedings of 2nd International Conference on Micro-Electronics, Electromagnetics and Telecommunications*, pp. 333–342. Springer, Singapore, 2018.

Surekha, B., and G. N. Swamy. "Security analysis of 'A novel copyright protection scheme using Visual Cryptography'." In *International Conference on Computer and Communications Technologies (ICCCT'14)*, pp. 1–5. IEEE, 2014.

Surekha, B., and G. N. Swamy. "Visual secret sharing based digital image watermarking." *International Journal of Computer Science Issues (IJCSI)* 9(3): 312–317, 2012.

Thakkar, Falgun, and Vinay Kumar Srivastava. "A particle swarm optimization and block-SVD-based watermarking for digital images." *Turkish Journal of Electrical Engineering & Computer Sciences* 25(4): 3273–3288, 2017.

Thanki, Rohit, Vedvyas Dwivedi, Komal Borisagar, and Surekha Borra. "A watermarking algorithm for multiple watermarks protection using SVD and compressive sensing." *Informatica* 41(4): 479–493, 2017.

Verma, Vibha, Vinay Kumar Srivastava, and Falgun Thakkar. "DWT-SVD based digital image watermarking using swarm intelligence." In *Electrical, Electronics, and Optimization Techniques (ICEEOT), International Conference*, pp. 3198–3203. IEEE, Chennai, India, 2016.

Wang, Yuh-Rau, Wei-Hung Lin, and Ling Yang. "An intelligent watermarking method based on particle swarm optimization." *Expert Systems with Applications* 38(7): 8024–8029, 2011.

Index

A

Accuracy, 44, 56, 73, 112, 138
Additive Noise, 8
Additive Watermarking, 10, 31–32, 34, 51, 116, 117, 119, 126
Altera, 113, 116, 118
Ambiguity, 18
Ant Colony Optimization (ACO), 73, 83–84, 86
Application Specific Integrated Circuits (ASIC), 108
Artificial Neural Networks (ANN), 45, 48, 53, 63
Attacks, 3, 7–9, 13–19, 25, 27, 28, 32–33, 37–39, 41, 45, 49–55, 66, 77, 79, 92, 94, 96, 99, 107, 117, 135, 126, 135, 137, 139, 149, 153, 154

B

Bacterial Foraging, 73, 83
Bee Algorithm, 73, 83–85, 101, 106
Broadcast Monitoring, 8, 127, 134, 135

C

C/C++ compiler, 1, 122, 125
Capacity, 8, 14, 27–29, 44, 62, 64, 69, 84, 85, 90, 117, 138, 139, 140, 144, 154
Cat Swarm Optimization, 74, 83
Checkmark, 17
CMOS, 113, 118–119, 122–123, 130
Color, 5, 9, 22, 29, 33, 35, 39, 50–54, 57, 59, 61–62, 64–65, 67–68, 90, 101, 104, 117–118, 122, 125, 127, 140–142, 154, 155
Compression, 7–9, 13, 17, 28, 30, 37, 50, 52, 54–55, 105, 138
Compressive Sensing, 10, 22, 40, 58, 65, 68, 70, 141, 154, 156
Computational complexity, 14, 40, 140
Convolutional Neural Networks (CNN), 45–47
Copy, 2, 3, 4, 8, 18, 26, 134–136
Copyright Protection, 5, 8, 9, 23, 30, 35, 57, 60, 63, 104, 128, 134, 138, 139, 142, 143, 155, 156
Cryptographic, 3, 13, 17–18, 136
Cuckoo Search Algorithm, 74, 83, 85, 99
Customized IC, 113
Cyclone, 113, 118, 122

D

Data Authentication, 23, 59, 135
Data Payload, 14
Deep Learning (DL), 10, 47, 56, 71, 140
Difference of Gaussians (DoG), 28
Differential Evolution, 73, 80, 99–101, 105

Digital Imaging and Communications in Medicine (DICOM), 136
Digital Millennium Copyright Act (DMCA), 2
Digital Right Management (DRM), 1–3, 134
Digital Signal Processors (DSPs), 108
Disable Detection, 18
Discrete Cosine Transform (DCT), 9, 27, 34, 78, 110
Discrete Fourier Transform (DFT), 27, 110
Discrete Wavelet Transform (DWT), 9, 27, 37, 78, 84, 110

E

Electronic Patient Records (EPR), 136
Embedder, 11–12, 26, 129
Expectation Maximization (EM), 28
Extractor, 12, 26
Extreme Learning Machine (ELM), 45

F

False Positive Rate, 14, 18, 56
Feature Extraction, 28, 46
Fidelity, 13
Field Programmable Gate Arrays (FPGAs), 108, 109
Filtering, 8, 17, 28, 30, 37–38, 49–52, 54–55, 138, 153, 154
Fingerprinting, 8, 134, 135
Firefly Algorithm, 74, 83, 90, 101, 102, 104–105, 106
Fitness Function, 75–79, 81, 82, 85–96, 147, 149
Fragile, 7, 29, 33, 37, 38, 118–125, 130, 135–136, 154
Fuzzy Logic, 45, 49, 54, 65, 71, 103

G

Gain Factor, 10, 32, 33, 39, 71, 85, 117, 143–144, 146–149, 151, 152
Gaussian Mixture Model (GMM), 28
Genetic Algorithm, 70, 73–74, 76, 80, 100–104, 141
Geometrical, 17, 66

H

Hardware, 12, 14, 23, 57, 107–133
Hardware Description Language (HDL), 109
Hessenberg Matrix Factorization, 41–42
Hospital Information Systems (HIS), 136
Host Port Interface (HPI), 111
Hybrid, 7, 9, 10, 37–38, 61, 63–65, 100–103, 108, 116, 125–126, 137, 142, 155

I

Imperceptibility, 13–16, 27, 33–34, 37–38, 44, 46, 71, 77–78, 81, 90, 99, 117, 126, 138, 139, 143–144, 147, 149, 152
Industrial, 64, 66, 139
Integrity, 3–4, 7, 14, 22, 25, 131, 135–141, 154
Intellectual Property (IP), 1–2, 133, 139
Intelligent, 10, 19, 22–24, 45, 47, 58, 61, 64, 68–69, 102, 104–105, 128, 130, 132, 139–140, 142, 154–155, 156
Invisible, 7, 11–13, 19, 38, 49–55, 71, 118–125, 128, 130, 133, 135, 144, 147

L

Least Significant Bit (LSB), 7, 27, 29, 116
Legal, 14, 18
Linear, 8, 34, 39, 44, 46, 72, 117
Look up Tables (LUTs), 112

M

Machine Learning (ML), 7, 9–10, 19, 26, 44
MATLAB®, 110–111, 113–114, 118, 120, 122–123, 125–126, 131
Mean Square Error (MSE), 15
Military, 133–135, 137–138
Multiplicative Watermarking, 10

N

Neural Networks, 19, 44–45, 49, 61, 63, 68–69, 71, 99, 102, 104
Noise Visibility Function (NVF), 15, 49
Nonlinear, 8, 45–47, 72
Normalization, 28
Normalized Correlation (NC), 15, 81, 148

O

Optimal, 19, 45, 49, 50, 66, 71–72, 74–76, 79–81, 84, 86–88, 90, 93–94, 97–100, 103–105, 155
Optimark, 17, 21
Optimization, 9–10, 22, 45, 62, 72–75, 77, 80, 83–86, 88, 90, 95, 97, 99–106, 126, 143–145, 154–156
Oracle, 17–18
Owner Identification, 134

P

Particle Swarm Optimization, 74, 83, 88, 95, 100–103, 105, 155–156
Patchwork, 6, 29
Peak Signal to Noise Ratio (PSNR), 13, 108, 148
Picture Archiving and Communication Systems (PACS), 136
Predictive Coding, 30
Probability of Coincidence, 14
Protocol, 2, 17–18
Pseudo-Random Noise (PN), 117

Q

QR Decomposition, 40–41, 61–65, 67–68, 95
Quartus II, 113

R

Radial Basis Function (RBF), 45, 49
Rectified Linear Unit (ReLU), 46
Redundant Discrete Wavelet Transform (RDWT), 146
Region of Interest (ROI), 137, 140
Region of Non-Interest (RONI), 137–138, 140
Reliability, 14, 108
Remote Sensing, 134, 137–138
Removal, 13, 17
Resampling, 8, 13, 51
Rescaling, 8
Reversibility, 5, 14
Reversible, 8, 14, 42–43, 59, 62, 64, 102, 116–117, 121, 127–129, 132, 137–138, 139–140, 141–142, 155
Rotation, Scaling, and Translation (RST), 33, 49, 51

S

Schur Decomposition, 41–42, 59, 62–63, 65, 67
Sensitivity, 56
Simulated Annealing, 73–74, 97–98, 99, 102–105
Simulation, 57, 109–111, 113–116, 117, 124, 147–149
Simulink, 110, 114–115, 131
Singular Value Decomposition (SVD), 9, 38–39, 78, 81, 126
Sparse, 10, 40, 68
Spatial, 9, 26–30, 32, 37, 49–51, 53–54, 58, 67–68, 91–93, 96, 100, 108, 112, 114, 116, 126, 128, 131, 142, 154
Specificity, 56
Spread Spectrum (SS), 32, 116, 120
StirMark, 17, 21
Structural Similarity Index Measure (SSIM), 13, 79, 92
Support Vector Machines (SVM), 44, 71
Support Vector Regression (SVR), 45, 71

T

Tabu Search, 74, 98, 101–105
Tampering Detection, 65
Telemedicine, 67, 133–134, 136, 142
Texture Mapping, 30
Transaction Tracking, 26, 134–135

U

Universal Asynchronous Receiver/ Transmitter (UART), 115
Universal Multimedia Access (UMA), 2

V

Verilog HDL (Verilog), 109
Very High Speed Integrated Circuits (VHSIC) HDL (VHDL), 109, 118, 121–123
Virtex, 113, 116, 118–120, 122–123

W

Weighted PSNR (WPSNR), 15

X

Xilinx, 108, 110, 113–114, 118–125